TECHNO WITCH

Ancient Wisdom, Digital Tools

E.E. Warren

Technowitch: Ancient Wisdom, Digital Tools
by E. E. Warren

Cover Art: Digital Juice, design by Ellie Warren
Editor: Taylor Ellwood

Font: Palatino

ISBN: 978-1-7345786-2-1

Author website: www.homehearthmagic.com

Dedication

To Schoen and Missy, as always. Schoen helped me see that I could do this…and without him, would I have? "Finish the damn book!" Missy is my forever 'partner in crime'… (the 'she' to my 'nanigans!') and both of you have been so completely supportive. It means more than I can say; thank you, both of you, for your wonderful input. This barely scratches the surface. Thanks for the fun conversations leading up to this book, and I can't wait to write the vampire books with you!

To Jimmy, as always. You should be here for this, dammit.

To Robin Sojourner Wood: You unknowingly helped nurture my witchy side when it couldn't come out to play anyplace other than Second Life way back when. I never got the chance to tell you that chatting with you was something I would always treasure, as for some time, I had a serious case of hero worship, both as a wanna-be-creator in SL, and as an actual artist in RL. What a gracious, loving, caring soul you were, and what a sad loss for all of us. Hail the Traveler! Safe journeys, Robin. I will say your name often.

Finally, to all those people who allowed me to ask them questions, lurk around and observe, who gave me permission to use items they posted...thank you for your kindness and generosity.

Special thanks go to Rachel Capurso and
Andrei Burke, Chaweon Koo, Carolyn Elliott, and Taylor Ellwood. You all rock!

Table of Contents

Introduction by Taylor Ellwood

You are holding in your hands a groundbreaking book. I should know, because I edited it. I took this project on because I'm always intrigued by seeing pioneering work that pushes what we know about magic.

This book is a pioneering work. It's not a technomancer spell book, but what it is, is a holistic exploration of what technomancy is as well as what it can be. Ellie describes the possibilities and shows all of us what can happen if we're willing to stretch outside the bounds of what we think magic is, or what it ought to be.

Of course, the nature of a book like this is that it will it inevitably become outdated (in some ways). The technology will change and advance. But the principles and concepts and ideas and practices described within will be sound, even in a new technological environment.

I invite you to read this book with an open mind and challenge what you know so you can discover what you can learn.

Taylor Ellwood
Eugene, OR
September 2021

Why This? Why Now? Why Me?

First, some caveats. This isn't going to be the longest book you'll ever read.

It won't be the 'be all and end all' of books on the subject. I'm not an expert, and I'm not about to pretend to be one. It would be disingenuous of me, and to those who have been knowingly using tech in their practices longer, or if you're a technomancer, it would seem rather presumptuous.

I found myself researching technowitchery for a story in my upcoming fiction book, "The Magic Around Us." I discovered by and large that there isn't much that's light on the subject available, or meant for beginners - and considering how some topics have such a wide breadth of information, there truly wasn't much available on the subject in general. You have to know what you're looking for, be researching the topic on purpose, and then be able to dig what little bit you need out of a much larger work. That's fine. I suppose it both acts as a gate, protecting people from knowledge they may not be ready for, and also forces folks to actually DO the research. Researching and then turning around and experimenting with what has been learned is how witches work best anyway. That way, what we know is experiential, and not all book learning.

So, why would I think about writing a book on technowitchery if I'm not an expert?

It was something I realized as I was researching...and ultimately, it convinced me to write a book - this book - that you now read.

I started out researching, initially, by looking at articles that had been published, and on social media, and I asked questions in groups, hoping to find witches willing to discuss the topic with me, so that I could then glean from them the information I needed.

Well, that's what I planned on, anyway. All too often, what I **did** is ask a question, which started arguments. It was not what I meant to do, but that was all too often the result. Sorry if you ended up in one of them. Several were quite heated. Want to know what I learned?

Some witchy groups HATE the idea of tech mixed with magic, and argue that we need to stay more traditional in our workings, feeling that tech isn't reverent enough, or doesn't show enough respect to tradition. Other witchy groups LOVE the addition of tech, and counter back with the idea that witches are nothing but resourceful, and what is a crock pot anyway, but an electric cauldron? They argue that our esteemed ancestors would've chucked their cauldrons for a crock pot (not to mention electricity!) in a hot heartbeat, and never once looked back. If you're on social media in witchy groups, these polar opposites aren't...unexpected. The depth of feeling and angst toward it might be. Believe it or not, that exact topic - crock pot vs. cauldron - was one of the most heated arguments.

While I was in the middle of starting arguments with random lovely witches, hanging there in the middle of the tech vs. tradition argument is that 800 lb. gorilla in the room that became immediately apparent to me, once I started researching the topic of technowitchery. There are people already using forms of technowitchery and are engaged with it because of the very nature of social media and pop culture, and...surprise!...some of them are the very folks who argued against it! Some witches have been witches long enough that they may realize what they're dealing with when they see technomagic on social media. However, for every witch that knows what they are dealing with, there are many more who don't. If you're a new-ish witch, you might not realize it when you see it. If your life runs at what feels like a hundred miles an hour, you'd probably pay more attention and recognize it, if you only had the time. Maybe you view things in your life compartmentally - there's a time and place for witchery, and then social media is how you catch up with friends and family. Some folks, like me, just weren't paying that much attention. Like, scroll, scroll, love, comment... You get the picture.

Anyone who's been a witch long enough knows they shouldn't use magic they don't understand, or experiment with magic unless they're fully prepared to take some precautions, as well as deal with the fallout of their experiments. THAT is ultimately the reason why I decided to write this book. People are using forms of magic that don't even realize count as technomagic. I'll be the first to tell you **I was one of them.** Once I started researching, I realized that there were many, many ways I already was using technomagic...without ever thinking twice about it. I was guilty of using magic I didn't understand. Now, I do, and my actions are more informed.

This book is meant to educate folks in a light-hearted, easy-to-comprehend way. People can use what they already engage with magically in a more knowledgeable manner. Maybe it'll peak their interest, and cause them to go research other topics that appeal to them, and it will help them avoid pitfalls that had never occurred to them. It's not meant to be a how-to technowitch spell book. It's merely an introduction to the subject. Maybe you'll even learn you've already been practicing technowitchery yourself.

Before we start exploring what's hanging out in cyberspace, let's set up our first boundary.

What *IS* technomagic, which some also know as technomancy? What is a technowitch? There's actually a bit of confusion in the terminology. Some people use the terms technomagic and technomancy interchangeably, but I don't find them to be interchangeable. I'm sure every person has their own definitions, but here are mine:

Technomagic is when modern technology currently available to many people is used in unexpected ways to perform magic. Some of these 'unexpected ways' to use tech may be quite different than their original intended

you feel shouldn't have it; Fear of criticism, or, fear of ridicule, either within your community, or outside of it, if suddenly the people around you realize what you're doing. Whatever your fears, this book is not meant to be an in-depth, definitive treatise on technomancy practices, or be the written equivalent of giving razor blades to toddlers. I know most of you keep a low profile on purpose, and are firm believers in the statement "To know, to dare, to will and to keep silent." Please know I heard you - but my fears stem less from people learning new magic, than people unwittingly using magic they don't realize they are using.

Other groups, similarly, may find subjects or topics they disagree with. Some will disagree with various topics in their entirety – like Pop Culture Magick – or see certain kinds of practices, or even places where younger, and/or newer witches gather - such as TikTok or Tumblr - as invalid, and question those that frequent them as being seen as witches at *all*. I would ask everyone to try and keep an open mind. There is no gatekeeping here. All paths are valid.

And to those who fear the blending of tech and magic, think about this: When electricity first began to reach most houses, how many people didn't trust electricity in general, or lights in particular, for fear of being electrocuted? Thomas Edison and Nikola Tesla had a "war of currents" that had Edison showing grotesque imagery of animals being electrocuted by means of Tesla's alternating electrical current that was meant to make people afraid.[1] After the vacuum cleaner was invented, how many people were afraid of them, and continued to use a broom? A hundred years down the line, electricity, light bulbs and vacuums are considered necessary, if not essential to modern life. It wasn't always that way. It doesn't make candles or brooms wrong. It doesn't do away with them; it gives you one more tool. The best tool is the one you use. A hundred years from now, will those who follow us look back and wonder what we were so concerned about?

I would add this final note: Know that anything I shared within these pages, I got *permission* first - not only to share their practice, but whatever name they wanted associated it, or no name, if they chose.

Last thoughts before we dive in:

Coming up in the chapters ahead, I'll tell you how I discovered I was using technomagic, and then move on to some of what I found as I researched it that you may find helpful in your own path, or recognize as something you already do. I discovered fairly quickly that anything witchy you can do or that works in real life, you can successfully do in the virtual realm of

[1] https://www.history.com/news/what-was-the-war-of-the-currents

cyberspace. Some of them may never have occurred to you to do, much less consider it to fall under the category of "technomagic." We witches are, after all, a resourceful lot.

Remember that there's something here for everyone, whether you're new to magic in general, or you've been at it forever...or you love your cellphone, but aren't on speaking terms with your computer. No matter what you feel like you do or don't know about a topic (whether magical OR tech-related) don't be afraid to seek out more knowledge. Everybody starts somewhere! To help bridge the gaps and give you a start, there are several appendices at the end, including a listing of witchy resources that are available for the downloading (some pay, some free), plus a listing of some of my own resources - books and articles - that I used or found interesting as I wrote this. That way, if you decide to dive a little deeper, I've provided your first stepping stones. Additionally, there's a photo gallery organized by chapter that contains any pictures that were referred to in the text.

Of Sacred Spaces and Places

Sacred space.

What is it to you? What does it mean to you? What does it feel like to you?

I ask because we don't question it in real life. We step into a sacred space: a temple, church, or simply in front of an altar in your own home, and we don't think about it. It's there, you're there, and you interact with it. You can palpably feel it. For me, there's a quietness to my soul, and a reverence in my heart, as well as a connection to something magnificently beyond myself that reaches back down to me.

How do you get that same feeling in a non-physical space that doesn't actually exist? **Can** you? Of course you can! However, many people assume that it's impossible to achieve the same range of feeling in a digital space. It's one of their arguments that virtual spaces can't or won't ever compare to "real" sacred spaces. Other arguments I've heard are that deities don't like it; it's not reverent enough, and it's too non-traditional.

First, you have to believe you CAN have that same feeling. Intention is everything in all of magic, but it's paramount when you speak in terms of technology and magic. Whether it's setting up your sacred space or creating a virtual altar, to doing spells, intent creates and powers it all. If you don't believe it will work, guess what? It WON'T (which is the same as traditional magic.) Futurist and technopagan Mark Pesce noted back in 1995 in an interview for WIRED magazine that:

> *"Both cyberspace and magical space are purely manifest in the imagination. Both spaces are entirely constructed by your thoughts and beliefs."*[2]

If any spaces we use magically (virtual or otherwise) are purely in our imaginations, it would seem to me that whether or not they are sacred has everything to do with our beliefs and intent. That is the first thing you need to tackle: to make up your mind that the digital counterparts to your normal places, spaces and tools are just as valid as the physical ones…(or at least allow that it could be.)

The next step is to decide the specifics of your sacred space, whether you commune with, worship or work with a deity - or not. Some people use their sacred spaces or their altar space specifically dedicated to a purpose (like worship and offerings), others do spell work at it. In technomagic, you should work within a container - a space that if nothing else, is sacred to YOU. Where will it be? What would it consist of? What do you need? Depending on what you plan on doing at your altar, it may determine the type of tech you use to create it.

---

[2]https://www.wired.com/1995/07/technopagans/

All witches are different in how they work. There are some folks that follow a spell like a recipe; others make their own up on the fly. Still more will take a spell someone else created off the web or from a book, and modify it to their own needs. Others see spells as 'prayers with props' and rely exclusively on intent, with little ritual involved. It's driven purely by personal preference. None of these are wrong; all are valid. The important thing is knowing yourself, what you prefer to do, along with what works for you and feels right.

Knowing how you work will help you determine what form your sacred space should take. There are almost as many ways to have a virtual sacred space as there are people, and there's lots of talk about it currently, due to COVID and social distancing. For instance, are you a solitary witch, or do you belong to a coven, or do magic with your friends? What is your level of expertise with tech? What a coder or programmer would do with sacred space might be considerably different than someone who doesn't code or program.

The easiest way to carve out the simplest of virtual sacred spaces is to create a photo gallery online that is specifically intended for that purpose. You'll laugh, but I think every witch should have one, and all it needs to have within it is pictures that bring the sacred to mind for you. It sounds almost TOO simple, but it's actually quite effective. I'll use my own as an example:

I am Sekhmet's, in regard to deities. (That's me; I know not everyone reading this works with deities. You do you!) My most convenient virtual sacred space is my photo gallery, which resides in a folder on my Google drive that only I can access. Inside that folder, I have photos of my physical altar, as well as photos of Sekhmet and other meaningful photos and music.†[3] The same photos/music are stored as files directly on my cellphone (not only in the cloud). I have it in both places because not all areas have good internet signals, and what if you ended up someplace without a signal?

As I write this, I'm sitting in a car service center, iPad in hand, waiting for my serpentine belt to be replaced. The three other folks waiting (socially-distanced and masked) are all glued to their cell phones. If I add earbuds and go to my gallery, I can be in my sacred space. I have a signal, but if I didn't, I could still access it, and enjoy it. Is my sacred space any less valid if I'm not physically there, but my mind is? Does Sekhmet require my physical presence, or my mental attention? What does any deity crave but attention and devotion? Even if you don't work with a deity, being able to access a sacred space in the middle of a hectic afternoon can be a blessing.

Another way to carve out some virtual sacred space is to have a digital temple.†[4] It can be as simple or as elaborate as you choose. You can create a webpage for free. Most of the places that offer free web pages have editing tools or templates available that make designing them easier. Once you decorate it with meaningful items, such as photos of your physical altar, or someplace that holds sacred meaning to you, you can

---

[3] See Photo Gallery, picture 1
[4] See Photo Gallery, picture 2

access it from your cellphone or a computer. It can be accessible to as many people as you choose: open to the public, only to a few select friends, or coven mates, or to you alone. If your online temple is shared with others, many have chat features that you can type in real time to one another, if you choose. There are also ways to use Discord, Skype and Zoom if you want to work with others and be able to speak to them in real-time. In the age of COVID, Skype and Zoom are now familiar to most. Discord is routinely used by gamers playing World of Warcraft, Minecraft, Second Life or any other number of online games that have multiple players. All of them facilitate people getting together digitally, regardless of the purpose that gets them together. The only requirement is that whatever you choose suits your needs. There's no right or wrong answer as to what it should have. You can make it interactive and be with others corporately, or simply create something beautiful that puts you in the right frame of mind.

My personal favorite place for online sacred space is within a game called Second Life. Why? While Second Life is called a game, and there are people playing games within it, it's totally freeform, and in that is truly a bit like life. Linden Labs, the developer of Second Life, doesn't dictate what it is or should be. There's no plan. It's as individual as the people who populate it. The main rule in SL is the Golden Rule (also known as, "Don't be an ass"). Whoever you are, however you identify, whatever you're into - you'll find others who share your interests.†[5] I have an avatar who embodies my spirit in a digital world. I have a digital home, which is also my sacred space. It is conducive to meditating, even in real life. There are wind chimes. I have an altar. I have a fire pit, and twinkly lights. It is peaceful. This suits me. Meanwhile in Second Life, there are people going to school; there are companies with offices in far-flung places that use it to have virtual meetings. There are also various interest groups. Want to be a vampire? A fairy? How about a mermaid, or a werewolf? Want to spend time on a Federation space station? They're all here. Like different eras? The past? The future? Steampunk? Pretty much every interest you could imagine is represented - including witches. It's only limited by your imagination. If you're a social creature, you'll relish the ability to find other like-minded individuals to do rituals with. Introverts and folks who are solitary practitioners can find the peace and quiet they crave. If you're a gamer, or you're good with tech, be sure to check out Second Life. (There's a whole chapter on the intersection of gaming and technomagic in Chapter Six.) Some folks might find some of the actual 'mechanics' of having an avatar in a virtual world to be cumbersome, or daunting, but there are lots of friendly folks who are willing to help you.

How Do I Make My Own Sacred Space?

Now that we've discovered some forms sacred

---

[5] See Photo Gallery, pictures 15 & 16

space can take, how do we actually create it?

Like your actual preferences on how you work magically, the way you create sacred space depends on what you want as an outcome. Again, remember: it's about your mind. When you're in a virtual space - whatever form that may take - your mind doesn't differentiate. It's just one more place you went to, even if your body didn't follow. Whether it's spending time in your virtual sacred space calming your mind, or spending time with your deity of choice, or casting spells, it doesn't matter. Your mind doesn't care. It sees it as real, and it's just as legit as if you were physically in your sacred space.

The simple act of deliberately creating a new folder on your computer or a flash drive to put pictures into can become a magical act, as can deliberately moving the appropriate photos into it, particularly since you just created them into existence. No matter what, note the term "deliberate." Make no mistake. The creation of folders and moving of files, in this case, is very much a magical act. Some folks may wish to do a short spell, or offer words of dedication. For many, the intent alone would be enough. Do what feels appropriate for you. Ask yourself, which piece of tech do you use the most? For most people, it's going to be the one you carry around with you - a tablet or cellphone.

First, take your device, and smudge it. Use whatever tools you normally would use for smudging. I use palo santo sticks, or incense if I'm out of palo santo. Before you start to carve out a sacred space or use it specifically with intent as a tool in witchery, you'll want to smudge it to cleanse it, and get the 'digital grime' off of it. Another way I've heard is to leave several crystals on the screen, with the device turned off. If you know or use another way that you'd rather choose, go for it! Either way, other people's emotions need not hang around to cloud your own magic, lessen the efficacy of your spellwork, or infiltrate your sacred space - particularly if you're frequently on social media on whatever device you use. See any piece of tech as a magical tool, and treat it accordingly!

Once you've smudged your device, you need to consecrate it in whatever way feels comfortable for you; That's one way. Feel free to use it! If you have your own idea, do that. This is where the "There are many ways to cast a spell' part comes in, or simply just your intentions.

Next, you'll want to actually create a folder.

After you've created your folder, you'll likely want to dedicate your device, the folder your items will reside in and the sacred space you're creating to your best and highest good (whether it's living directly on your phone or on the cloud somewhere.)

At this point, you have an option to make it stronger in your mind that the space is set apart as sacred. If you think you need the reminder that it's a sacred space, you could type your dedication into email or a notepad, or handwrite it and take a picture, and then send it to yourself in email or copy it into your folder unopened. Seeing the unopened file every time you use your space reminds you that it is a consecrated space.

How and where you create your space is entirely up to you. The important thing is that you know that you can create your sacred space. Now, put items into it that MEAN something to you. Something that gives you lovely, gooey feels, that makes you smile, or evokes feelings of thankfulness, reverence, joy...hope. Whether that's photos of your physical altar at home, a deity, your lover, your kids, music that puts you in a specific frame of mind, or photos that remind you of whatever symbolically says "sacred" to you - put them in that folder...and enjoy it, particularly on days that they seem a little further away or harder to hold. Don't limit yourself to photos! I don't know about you, but for me, the music I store in my sacred space folder is important. I've got several folders of tunes; one that puts me in a reverent frame of mind, another that feels 'witchy', and a final folder that has music specifically meant to either pick me up on a crappy day, or have me dancing at my desk or in my kitchen for the sheer joy it evokes within me on a good one. Seriously, go to town with what you put in your folders. Not only does it get you in front of an altar more often, it's a little bit of self-care, too. Win-win.

How elaborate your virtual sacred space is...and including what kind of technology it uses, where it resides and how you access it - that's entirely up to you. A folder on your phone is likely to be completely private. A folder on the cloud, however, can have shared access. Do you want your space all to yourself, or do you want to allow others to see it? Or even have others participate by adding their OWN folders (a little like each of you having a "shrine" in a larger "building" that you can freely visit between.)

All those ways I mentioned earlier are all valid ways of having a virtual sacred space, and there's half a chance as you've read this, you might think of other ways that I haven't mentioned. Whatever you create, with the intent for it to be sacred, it becomes a sacred digital space you can work with.

Some of the ways to create virtual sacred spaces get more technologically challenging as we move forward. For instance, creating a webpage takes a little more technical know-how, and may entail a learning curve if you've never done it, though most free and pay web page hosts have creation tools that will walk you through the process. In the same way that creating a "new folder" to put your pictures of what's sacred to you is a magical act, so is creating your webpage! What you choose to put on your page is entirely up to you. What would **you** put on a webpage that serves as your sacred space? All of my virtual sacred spaces at least include a picture of my physical altar, plus some music and wind chimes, but that's me. What about you? What calls to you? Imagine you're having a bad day at work. How wonderful would it be if you could go to a simple webpage that you've set aside as sacred? Is it any less a refuge? (See the 'tools and helps' appendix for links to examples of a web temple.)

If you're a gamer, your ears perked up when I mentioned the use of games as a 'virtual sacred space,' and you're probably already either trying to wrap your head around how you could possibly use a game as sacred space, or you're imagining ways you could use your favorite games magically - or you soon will be!

Second Life is my favorite place for magic, but nearly any game has something you can do with it magically. In World of Warcraft, for instance, there are places I take my avatars that I find peaceful. Usually, these are out of the way places that are beautiful and evoke a proper feeling. Whatever games you play, there are always places that have less people. Think 'avoid current content.' I tend to avoid the 'newbie' zones, too, as they also tend to have more people. The places I find are usually in the middle of nowhere. It's the digital version of taking a walk in the woods. In the commercial version of Warcraft, if you need an altar, you have to go find one, because you can't make one, and you always run the risk of getting trolled or ganked by players, or possibly killed by non-player characters. Another option for Warcraft players is to have your own private server. From what I understand, aside from Warcraft, there are many other games that allow you to run or join a private server. If you play MMORPG games and have a favorite - even if it's long gone - do yourself a favor and check it out. There may be an opportunity for you to start up a private server, or there might already be one. I know Final Fantasy has private servers, and multiple 'dead' games are still being played via private servers. The benefit to a private server is that when you have your own, you can choose to let other people in - or not. Some private servers, like Warcraft, will allow you to not only play the game, but may give you the ability of an admin - to either rezz any objects from their catalog of items, or possibly create your own game content. It would depend on the game, on the server, and how they have it set up. If you want an entire world to yourself, you can have it.

What's Right for Me?

How you answer that question depends on what you need, truthfully. Do you work solitary, or in a group? Do you want to do this on your own, or in the company of friends? If you're working with friends, what's your 'common tech denominator'? Do all of you have cellphones, but several of you work behind firewalls, or can't have a phone at work, but are allowed to be on the internet? Know everyone's limitations, and go from there. What's its' purpose? Is it merely a place to remind yourself of the sacred, or are you meaning to do magic and spells? To rightly decide what kinds of sacred spaces you need, you need to know what's sacred to you, who you'll be with, what you want to do with it, and finally, apply all that knowledge to the technology in question.

You may actually want sacred space in as many of these forms as you can manage. Why? Think of each way you can have a virtual sacred space as a portal of sorts. All of them are doorways that lead to someplace that's personally sacred to you. You might be able to access a web page at work easily enough, but most employers take a dim view of game access, or frown on obvious phone usage. Some folks can't have their phones at work at all; others can have their phones, but the internet is locked down to pages that only pertain to their job, or some webpages are behind firewalls. At home, a game-based version might be perfect. If you find yourself traveling, your phone or a webpage might be just what you need. No matter what you choose, the more options you have, the more likely you will use it not only when you need it, but also on a daily basis, forming the cornerstone of a daily practice. For some, a daily practice is the gold standard, but hard to achieve. Virtual sacred space makes it more achievable. The best sacred space is the one that gets used consistently

Altars and Symbolism

Now that we've defined what a virtual sacred space can be like...What symbols are on your altar? We did cover it a bit in the last chapter, but now we get to be a little more specific.

Altars are known for being as individual as the people that set them up, and oftentimes, you have more than one. Go on Pinterest and search for "witch" and "altar," and you'll see very quickly how individual they really are. Sometimes they're more permanent, other times, they're temporary and are only there for a season, while you work through an issue (like finding a job or a new home) or work with a new deity. Virtual altars are the same. What is the power that lies inherently in an altar? What makes it sacred? Is it the sacred focal point it provides? Is it sacred because of what you have on it? Or is it sacred because you've worshipped at it, and your worship has consecrated it? Whatever answer you choose, not only is it not wrong, it holds true virtually as well.

On my physical altar, you'll find statues of Sekhmet and Thoth, as well as an incense burner, a candle, a small alabaster offering bowl, and a number of stones and crystals.[†6] There's a beautiful blue runner that everything sits on, and a particular stone that is my 'altar rock,' named Louie-Baba-Booey - Louie, for short.[†7] Yes, I really have a rock named Louie. I love that rock! He's

---

[6] See Photo Gallery, picture 3
[7] See Photo Gallery, picture 6

been part of every spell/working I've done, and he can either be found on my altar, or on my person.

When I'm away from literally everything else, he's a stand-in for my altar. I only mention Louie because everything we deal in as witches is underpinned by symbolism. Altars are no different. Altars are basically exercises in symbolism. Symbolism is the currency our brains deal with. Our brains are literally hard-wired to use them as shorthand for more complex associations that lead to emotions, which lead to intentions...and ultimately, our power, as they're all connected together. Louie, to my brain, is symbolic of my entire altar and everything on it, everything magical I've ever done, and I carry him with me. He stands in for my altar when I can't be there in front of it. I've done magic once or twice in a hotel room, and Louie's presence makes it feel legit. If a rock can be symbolic of my altar, why not a web page? Or the contents of a folder? Or elements in a game? Symbolism is symbolism, regardless of the form it takes.

Rachel Capurso, owner of Aeolian Heart Astronomy, had a wonderful definition of symbolism recently, in her class called CASHFLOW:

> *"The nature of the mind is metaphorical...the mind uses symbols, models and maps to understand the nature of reality. Your perception of the world is a map of reality written in neural patterns...revealing a myriad of associations and correlations between imagistic symbols*

and external reality."

Louic is an interesting, cool rock, with an awesome vibe - but ultimately, he's still a symbol that serves as a map in my mind: Louie = altar.

Carolyn Elliott, editor of Witch magazine, author of Existential Kink, and teacher said,

> *"Magic is an ongoing conversation between the conscious and unconscious parts of ourselves. The further I've gone down this path, the more I've seen that the "results" of magic are actually side-effects of using the language of symbols to speak to the great mysterious Other - (we could call that Other a lot of things - Goddess, God, the Universe, Nature, the Higher Self, etc.) - who then speaks back in terms of highly-charged, densely symbolic occurrences in real life, some of which "rhyme" with the conscious intent of a spell I've cast, others which "rhyme" with previously-unconscious intents.*

Whether virtual or physical, the trappings of our magical workings are symbolic. When you put objects on your altar, ask yourself why they're there. What do they mean symbolically to you, or are they there for aesthetic reasons (this one's pretty...that one's blue, etc.) There's nothing wrong with things being on your altar because they're pretty. If you're like me, when you're looking at other witches' altars on social media, it fills you with a sudden urge to redo your own! Beautiful things are a balm to our eyes, our minds and our souls. They have more power, however, if their symbology comes from deep within you. What do they mean symbolically to you? You may also find yourself asking, "What do these items mean symbolically to my patron deity?" and "What does that represent to me?" If, as Carolyn stated, magic is an ongoing conversation between the conscious and unconscious, and as Rachel said, that there are a myriad of associations and correlations between symbol and reality…then when you put items on altars...they're more. MORE. They aren't just pretty bits to honor a deity with, or to remind you of whatever the altar's purpose is. They're shorthand, reaching straight out of your depths, powering **all** of your magic and setting fire to your intent. Know your symbolism. Use it with intent and purpose. Know what powers your conscious mind (and gives you yummy feels) and know or learn what powers your unconscious mind, which may end up quite a bit darker as you plumb deeper into it, but it will be certain to contain charged emotions. (If you want to delve deeper into learning what your 'dark side' is, also referred to as 'shadow work,' check out Carolyn's book, "Existential Kink.")

The cool thing about symbolism and its effect on EVERYTHING magical that we do is that it doesn't care whether they're physical or virtual: the trappings of our workings are symbolic. Symbolism is abstract to start with, and that's why some people have a harder time grasping why symbolism matters so much. And then, because it IS so abstract, what means something to one person doesn't necessarily mean the same exact thing to another. It becomes personal very quickly. Symbolism matters to what you put on your altar; symbolism matters to your spells and worship. Symbolism actually goes quite a bit further in the virtual realm. Why, you ask? Because symbolism is an abstract concept where the roots lie in the mind of each individual. It actually requires a little more effort to make sure the symbols in your MIND adequately get portrayed in the physical. It's a "near approximation." When you're dealing in abstracts mentally, within a virtual realm, it takes off the brakes.

What you do and how many altars you have needs to resonate with how you normally work. Do what works for you! I have a number of altars in various places, much the way as sacred space. I have four of them:

My Google drive sacred space holds photos and representations of what is sacred to me, which includes photos of my altar and what's on it. I use Google drive, but any cloud-based storage would work, so that it's available when it's needed regardless of device. It's usually my refuge for on-the-go stress relief.

I have Louie the altar rock, who either lives on my altar, or comes with me wherever I go.

I have a physical altar in my craft room. Most of my spell work is done there, or starts there -- ends up in the kitchen, or by my fireplace -- and then ends up back in front of the altar.

I have a virtual altar in my virtual home in Second Life. It's a pretty table in the loft of my virtual house, with spell candles, a functioning book of shadows, photos of family and statues of Sekhmet and Thoth.

Wind chimes complete the setting.[†8] At my home in SL, it feels the same as when I'm in front of my physical altar. I also have a smaller altar on my fireplace mantel in my virtual living room, that's specific to my writing. It's a peaceful place.

Wherever your altars are, whether they're virtual or physical - heed your symbolism, and make sure the objects on it mean something to you and have a strong emotional attachment. You'll be using language that lights a fire under your subconscious mind, which is delightfully powerful.

---

[8] See Photo Gallery, Pictures 4 & 5.

What of Our Tools?

Once you get past the point of sacred space and altars, the first thing that usually comes to folks' minds would be, "Well, what of my tools?" I can all but hear some of you say, "I have (name a tool) and it's awesome, but there isn't going to be a virtual equivalent that could ever take its' place. What do I do?"

What tools do **you** normally use in your real-life practice? What couldn't you live without?

Most of us have some sort of cards (whether they be tarot cards, or oracle cards), maybe runes. My sister prefers runes. I've always been partial to pendulums. Many use scrying mirrors, or crystal balls. Some have wands, or athames, or both. Most of us have candles; if we work with a deity, there will also be items on an altar that are sacred to them. Most of us have a Book of Shadows, or some kind of magical journals we keep.

If you're wondering if there are digital equivalents to our physical tools available, there are. The biggest ones we have are our cellphones, tablets, and other mobile devices, not to mention our computers. If we use them in magic - they're a tool. Some of our tools are items we have around us, used in quite different ways than we would expect. Most of them are apps for our phones. Of the apps, some of them are paid, and others are free…and you probably have some of them already on your phone, out of convenience. I'll discuss apps as tools in general here, but know that there is an appendix at the end that has a list of apps, with descriptions of what they are, and what they do.

So, let's dig in.

Our mobile devices provide us a way to get to virtual versions of our sacred space and our altar - which already makes it an important tool. Many of our other tools either will reside on our mobile devices, or be accessed through it.

My most important tool would probably be my Book of Shadows, and it might be yours, too. Most witches have a Book of Shadows in some form or another. Some take the form of a simple notebook, while others are quite elaborate, with pretty parchment; many are even hand-illustrated and written. You can even purchase pages, download them and print them out, and make it look like the Sanderson Sisters' book from Hocus Pocus, to the book from Charmed, or you can scrap and crop your way to a custom book. I was in a class with a talented friend – who actually drew/illustrated the class notes – complete with little line drawings of each of us – and gave them to us to keep. Whether it's purchased complete or is self-designed/illustrated/made…how it looks or the form it takes doesn't really matter, other than giving you good feels and vibes. All that matters ultimately is that whatever you DO works for you. What good is it to have it, but you either can't find what you're looking for, or don't bother using it?

True confessions time! I started out with the best of intentions when I realized I was a witch. I bought a blank book, and then a beautiful hand-tooled leather cover of the Tree of Life to put it in. I included spells, correspondences, dates of spells as they were cast, followed by the details of my results. I even included an index, which I somehow thought would help me be SO organized! Unfortunately for me, it ended up more like every diary I ever had - unfinished and incomplete, only WORSE... It also had lots of little scraps of paper tucked inside as I wrote spells on the fly, used them, dumped them in my purse, and then eventually tucked them inside for safekeeping. It wasn't ideal. At best, it was unorganized; at worst, totally ineffective. I actually LOST it in my house twice. Really.

One of my friends came to the rescue. She pulled a thumb drive out of her pocket. Big deal, right? I have a dozen or more of them, and since I have three cats, sometimes I find them in odd places.

It was, actually, a HUGE deal...for me, anyway. Maybe for you, too.

That thumb drive's sole purpose in life was to be my friend's Book of Shadows. She had folders for everything from correspondences, to a password-protected diary, to pictures of her spells, and screenshots of things she wanted to remember that could prove to be useful, not to mention pictures she'd taken of scraps of paper where she'd written things down. Yay me! Scraps of paper that don't fall out when you open up your book!

Be still, my beating heart.

My thumb drive is now what I live out of, and it

works for me.†[9] Is it pretty? Not so much. Elaborate? Nope, but it does work! Several friends now have variations of it, or have sought other digital tools. There are quite a few to choose from, and if this is an issue for you as it was for me, I'm pretty sure that you'll find something that works. One of my friends uses Evernote, which I like, and has the added bonus of being cross-platform, and updates across devices and it can be shared easily across a 'team.' I can access it from my computer, phone or tablet. Another uses Microsoft's 'One Note' (which after seeing how she was using it, I'm working on re-doing my BoS on One Note.) One of my friends is an artist, and creates beautiful collages (much like those fancy pages for sale on Etsy) in Photoshop, and then converts the pages into pdf files, stashed in folders named 'spells', 'correspondences' and the like. There are also numerous diary or journal apps that could be repurposed into a Book of Shadows if you find one that you really like. There are pre-made Book of Shadows apps, but to me, they mostly didn't seem personal, or symbolic enough for me. If you have one and like it, don't let me discourage you from using it. To each witch, their own. The best Book of Shadows? It's the one that gets used!

As I mentioned earlier, most tools have some sort of equivalent, usually in app form. There are literally apps for everything these days...from your Book of Shadows that I mentioned, to your form of divination, to an actual altar, to your candles. If you've got programming chops

---

[9] See Photo Gallery, picture 7.

and are a witch, you might consider thoughtfully and mindfully creating apps that other witches can use as they realize the power of their digital devices. Why not get paid for helping your fellow witches expand their practices? Witchy apps don't have to be silly, trivial things ; they can actually be serious tools, used for the same reasons you use your physical ones. Some apps I've seen *have* been created thoughtfully - and others seem like they were developed by a team, attempting to capitalize on the recent influx of witches. Because of this, and the fact that everybody's different and has different needs/preferences, do yourself a favor. When looking into witchy apps, try a free version first, or ask others in forums such as Amino, Reddit or your social media of preference for their opinion of it first. It might either save you time and money, or you may get recommendations for apps you've never heard of, but soon find you can't live without.

What of those tools that aren't apps? They aren't limited to your digital devices. Remember I mentioned that some of our tools are items we all likely have around us that are used in quite different ways than we would expect? Here's where creativity goes off the hook. I've heard of people using laser pointers as wands, and using KitchenAid mixers, microwaves, coffeepots and freezers in their practice, as well as scanners and fax machines! The sheer volume of smart appliances and objects in our homes tells me that somehow, somewhere, a creative witch is using the technology in a way the engineers and designers **NEVER** meant or ever contemplated it being used. How many of us with a smart fridge would be tempted to put spells on the virtual memo board on the front of it? My first thoughts are for the health and welfare of my family, or maybe a self-control spell to keep myself from snacking between meals! Or if you have smart door locks, find a way to work protection spells for your home into it? As always, use what appeals to you, and works. Better still, consider sharing your practices around. If you have smart objects in your home, and you're using them magically in a way other people aren't, why not tell people what you do, and how you use it? What works for you might not work for me...but it might work for another witch desperate for something you do or use on a routine basis.

How Do I Clean My Digital Tools?

Like any of your other tools, you'll probably want to consecrate them before you use them. And as you use them, there can be a buildup of energy, or you may feel a need to clean them. So how would you do that, anyway? You may think of other ways, but here's what I do: When I install an app, I take a moment to ask it to serve my best and highest good. You may want to do the same as you do with some of your other tools when you consecrate them. When it comes to cleaning them, I smudge my electronics with palo santo sticks, and after some thought, if I were using 'other things' as tools - say a KitchenAid mixer, an InstantPot or other smart appliances, they should probably be smudged/cleansed occasionally like any other tool, particularly if other people may be using them, and their energy or emotions could potentially hinder your spells.

Virtual Spell Exemplar

Let me set a groundwork here before venturing further: this chapter isn't a spellbook, and is not meant to be. There are some spells in here, and it is certainly not a definitive, or an exhaustive list. The meaning of this chapter is NOT to teach everyone how to do spells. Chances are, those reading this already know spells that are at an appropriate level for their knowledge of magic and how it works. The reason these spells are here in *this* section is meant to educate folks about the variety and types of spells that utilize technology, and to get you thinking about what you see around you. There are multiple kinds of virtual spells, and all of them can be effective. Some deities are especially 'digital friendly' - particularly Mercury (or alternatively, Thoth or Hermes, depending on whether your pantheon is Greek, Roman, or Egyptian...) This is particularly true of Mercury, who was said to have invented writing, language, and trade. Others that I know that are walking a virtual path don't seem to feel that their chosen deities care in the least.

The best-known type of virtual/digital spell that people have heard of is likely an "emoji spell." You've probably seen a few. You may have seen articles on them; I know I did, maybe about a year or so ago, though even after reading the article, the only thing I really took away from it was that there was a 'structure' to them, in much the same way many spells rhyme. There are also "save" spells, "click me" spells, art-based spells, and memetic spells...not to mention traditional ones done in a digital or virtual way. Mind you, these are my own way of categorizing them, either based on how they're created, or how they function. Nowhere have I been able in my research to find ways that they have already been categorized.

EMOJI SPELLS

Emoji spells usually (but not always) have a specific syntax to them, and use those cute little emoji's you see in your message app, and most social platforms, which is where this type of spell had its' beginnings. Say you want to cast a money spell. The traditional spell I normally use including a candle and a Mercury dime is simple, at best, and wrapped in rhyme: "Money, money, come to me; as I speak, so mote it be." If I take this spell, and re-create it as an emoji spell, it would look something like this:

Note that it begins and ends with the same sequence of emoji's. This serves the spell in the same way that rhyming does a traditional spell. Some people use the sparkle emoji's to denote it's a spell, others use the magic wand emoji, or the crystal ball. Pick what resonates with you. Any spell you want can be a series of emojis. The way they are cast is usually on social media, like Twitter, Instagram, TikTok or Tumblr, and will say "Like to charge (which is adding your intention), and "Post to cast". The cool thing about emoji spells is that they go far and wide, and unlike a spell you cast alone at home, it can be charged by as many people that see it and like it, and every time someone else shares it, it's adding power and intent to your own cast. Yes, simply hitting the 'like' button charges it, but if you want to add more power to your emoji spell - before you hit like, focus your intent on it. When you post, before you post, visualize your desired outcome prior to casting. Like any other magic, it's our intent that powers it.

SAVED SPELLS

What if you want to cast a spell, but you don't need or want everyone you know to know your business? In a traditional way, you might write a spell on a bay leaf, or a piece of paper, and then burn it. A saved spell is similar to that, and there's a number of variations of it. The first is simplest: type your spell with intent into WordPad or some other text editor, visualizing the outcome, and then save it to a specific place knowing that saving it is casting it. Put it someplace you'll see often; that way it reminds you that it's still working, and every time you see the file, you'll charge it again.

Another way of doing this is to type it to yourself in an email, and send it to yourself. The act of hitting the 'send' button is the same as casting it. If it makes it easier to find, create a separate email account for spells and send it there. With any type of "saved spell" don't open the email! The thought on this is that every time you see the unopened email, it reminds you of your spell and recharges it. Once the spell is complete, and you've received your desired result, delete it.

Another more technical take on saved spells, particularly if you're a programmer or know enough about coding is to create a file that runs when you start-up your computer that has your spell typed in and commented out. Alternatively, you can code a spell to launch when your computer scans for viruses (ya'll *DO* virus scan your devices, right???) Either way, the coded file has your spell contained within it, and you know that every time your computer starts up, or your virus scan runs, your spell is being cast. Still another variation of a saved spell is to type out your spell word for word, and then change the font to a symbol font. It doesn't matter what symbol font you use. The one I used was Bodoni Ornaments, but only because it was the first one I ran into. It needs to be something that appeals to you, and that you can work with. For example, that simple money spell goes from:

Money, money, come to me. As I speak, so mote it be.

To this:

Basically, you've done two things: you've put a simple encryption on your spell, so that most people won't have a clue what it is or what it says when they see it; you've also reduced your spell to images, symbols that not only aren't readable, but are reminiscent of sigils. In fact, to strengthen a spell done like this, you could create a sigil. You can use a sigil generator, or create it yourself. There are several ways to create sigils, but normally when I do this, the above spell would have all the vowels removed, and any duplicate consonants removed as well, leaving me with these letters:

M N C T S P K B

Now, turn those letters into symbols. Whether you use the actual letters, or you choose to find a way to combine the symbols from whatever symbol font you used is entirely up to you. Those consonants then become:

The next part is the fun part. The symbolic graphic elements that remain can be turned into a sigil that is ITSELF a symbol of the spell you cast, that you can use however you wish.

Keep in mind; this is one way to do it; how I was taught. As my friend Nik Strange said, "There's probably hundreds of different ways to make sigils, for the most part I don't really think there is a 'wrong way' in a concrete sense, but there are ways of making sigils that simply don't work for some people. This isn't the fault of the method or the practitioner, everybody's energy simply works a bit differently."

For more information on how to create sigils, there's a link in the "Helps" appendix to Nik, who has a great blog that, among other things, has great explanations of multiple ways to create sigils.

SHARE ME/MEMETIC SPELLS:

These spells are exactly what you think they are. It's a meme that you 'like to charge', and 'share to cast.' Sometimes, it might say, "Share me unless you want bad luck," or show a pile of cash, and say "Share me if you want to be rich."†[10] It will either have the intent of your spell as a picture or it might be written out. The

---

[10] See Photo Gallery, pictures 8-10

difference between a share me spell and a memetic spell is at least the share me spell outwardly says it's a spell. A memetic spell, or meme spell, works EXACTLY the same way as a share me spell, but nowhere on it does it say 'like to charge, share to cast.' It depends solely on the image and text itself and they're counting on the meme generating enough emotion that you feel like you have to share it - whether those emotions be love...or hate.

Those who created it don't care what motivates you to share it. They just want you to share it, and when you do share it, it's either like/hate to charge, and share to cast. (Reminder: for screen shots of these types of spells, see the images appendix.) Always remember that intent is where our power mostly comes from. Hate, love, and fear are all deeply felt emotions, and are extremely powerful, indeed. There is also another form of a memetic spell that's considerably more insidious, labeled 'memetic warfare.' What kind of spell would be called memetic warfare? I'm sure you've seen them.

Imagine a political meme that says something very controversial. It doesn't matter what political party it targets. You see it on social media, and it either makes your blood boil, or you agree wholeheartedly and can barely wait to share it.[†11] Even as you post the meme, you can instantly name three people that will jump on your page and argue about it, with you and each other, and you can guess who will share it, and who won't.

[11] See Photo Gallery, picture 12

Depending on how hot button the controversy is you might even lose a friend or two over it. Congratulations, you've participated in memetic warfare. I know I have, and didn't realize that I had. There will be more later on memetic warfare, and how the memes work, and how to recognize them when you see them.

ART-BASED SPELLS

These are my favorites, actually. While you are creating a piece of artwork - ANY artwork - whether traditionally drawn or painted, or a collage of photos pieced together, then photographed and uploaded, or digital artwork or photography) you hold the intent in mind as you create your piece. Your intent, emotions, desire and magic are all mixed with the medium you're working with as you create, just like you would mix colors on a palette. You're mixing the magic right in.†[12] If you're really slick, you can add an actual spell to it subliminally so your image does double-duty magic work. For example, in a painting, painting a sigil in the color you're about to use first; throw your intent into it, and then paint over it. In a digital painting/photo, make a sigil smaller and/or partially transparent, and hide it in plain sight into your image. Some may see, most won't notice. Those of you who are artists or crafters of some type will be sure to think of other ways to incorporate your magic into your pieces beyond these. I once wrote out an entire spell, and then shrunk it in size to a dot, and made it a background

[12] See Photo Gallery, picture 11

for my computer at work. Anyone else would see a white dot and never question it. Me? I saw a spell in plain sight, every time I turned my computer on in the morning. That's the joy of being creative.

A Caution:

Now that you've seen the kinds of virtual spells that can be cast, I'm certain that you've seen most of them, if not ALL of them, long before picking up this book - without realizing exactly what they were. I know I did! This brings me to a topic to consider whenever you're cruising social media. Don't forget that memes (even non-magic ones, and ALL digital content) come complete with intent and very often play on symbolism and/or stereotypes that are hardwired into our brains - remember, they can give us power, right? They're maps, straight to our brains. Now, take a long, critical look at the memes on social media, with a fresh eye.

(Literally, go take a break, grab your favorite beverage, and hop on social media for a few minutes. Look at what pops up as you read.)

The memes on the social media of your choice may be using your power, tricking you into putting intent into something you might not. Or worse, it's playing on symbolism, or stereotypes in a thinly-veiled attempt to change how you think about something. Now, go to your own page, wall or posted content, and take a fresh look at what you've shared recently. Anything there give you pause? When I did this exercise myself - I was stunned. Whether you realize it or not, sharing something on social media, however well-intended - could be the equivalent of casting a spell, particularly if the content evokes high emotion. Symbolism is hard-wired into us. Intent and desire are what fuels our spells. Add charged emotions on top of that. In my brain, the equation looks something like this:

Personal symbolism + charged emotions* x *(intent + desire) = raw power.

On one end of the spectrum, there are many people who think magic isn't real. On the opposite end of the spectrum, there are those self-righteous folks that will suggest because you practice magic, or are some flavor of witch, you've got a one-way ticket to hell. What do both groups of people have in common? These same folks get on social media and have no clue that they may be making or using their own forms of magic. They're creating and using it without realizing what they're doing. There is wild magic run amok out there! All you have to do is look for it. Magically, social media is the modern equivalent of the Wild West. If today's world seems particularly chaotic, and somehow it always seems to circle back to social media...you just might be right.

Knowing what you do and WHY while online and on social media at least helps your magic to be focused. That's not to say that every meme you see is magic. Sometimes a meme is just...a meme. The point is that we need to look critically at what we see. The first knee jerk reaction may be to throw social media overboard; For some, that might be best. For most, however, we need to remain cognizant of what we're sharing. Before we share that funny or super-extra-opinionated meme - we should each ask ourselves whether it's something we would want to add power to, or bring about, or who it might ultimately benefit. Just as our words and what we say matters, so too, do the things we share on social media - which after all, are nothing more than pure intent and ideas, broken down into symbolism, which our brains simultaneously adore and are hard-wired for.

When Meme Spells 'GO DARK'...Or Memetic Warfare

Okay, so here's the current situation: There are spells out there on social media that many, if not most of the folks who post them don't realize are spells, and they may be emotionally or mentally tied to or invested in what the memes say. Beyond that, some of these same people don't even BELIEVE in magic, but they DO believe what's contained in the memes they forward. They act with full intent, without realizing what they are doing, or what they're utilizing. You end up with people using magic they don't understand, that they might not be happy to define as magic. That, in and of itself, isn't a good situation.

It gets worse when you add this bit in: there are people out there in that great digital, magical Wild West who knowingly and PURPOSELY use the general lack of knowledge people - magically-inclined or otherwise - have about online spells, the way information flows, and general human behavior and use them to the detriment of all of us by helping to turn the world into a chaotic mess, one 'funny' meme at a time.

An article by Vice Magazine from March 2017 entitled "Anti-Trump Witches and 4Chan Magicians Are Battling Over the Future of America" featured an interview with someone identified as "ATL"

> *"If it's seen as a joke, all the better," he replied. "A memetic magician's power is bolstered by laughter. Humor is a high-energy wavelength."*[13]

This particular article has a lot to say about Pepe the Frog, and Kek, and how they were involved in the 2016 elections. On one hand, you should know about Pepe and Kek, but I also don't believe this is the time or place to discuss politics. Suffice it to say that Pepe the Frog was used in many, many political memes. ATL, quoted above, had this to say about Pepe from the same article:

> *"When a lot of people pool their united willpower towards a single sigil, it's called a hypersigil, and it's exponentially more potent... Whether they did this consciously or not, it's exactly what has happened."*

This is one more thing that makes memetic warfare the scourge that it is. In addition to all the OTHER things that have been said about social media, such as people only tend to read headlines and blurbs, or listen to short

[13]https://www.vice.com/en/article/ywmp4v/anti-trump-witches-and-4chan-magicians-are-battling-over-the-future-of-america

soundbites that have been edited out of context, and usually ignore the need to fact check before they share items, there are folks whose jobs are to create 'funny' memes that poke 'fun' at particular people, groups, and movements, looking for tensions that already exist, and exploiting them. It's basic social marketing, and it's relatively inexpensive.[14] These people then set the memes in motion, and let those of us on social media do the rest of their dirty work for them.

We are, as innately magical beings, at an important place in history: we get to make history, as well as define it for those who will follow in our paths. We are the ones who decide how we will respond both individually and corporately to these kinds of magics. Do we use it, and let the chips fall where they may, and those who are not cognizant of what is swirling around them are on their own...particularly if they remind you on the regular that hell awaits you, or do we educate and enlighten any who would listen - whether they believe at all, or believe differently than we do about what havoc may be occurring unwittingly in their own lives, at their own hands or those of others?

As for me, I want to teach. That was my final, ultimate reason I decided to write this. You do you. No judgement here, and hopefully, you'll extend that lack of judgement towards me as well, if I reach out to those of different faiths to try to put what's happening in terms they'll understand. It may backfire spectacularly,

---

[13]https://www.heinz.cmu.edu/media/2018/October/troll-farms-and-fake-news-social-media-weaponization

but I feel like I need to try. The world's a f*****g mess, and I've always been the kind of person who when I see a problem, I don't just complain - I try to bring a solution. Remember when I said this was like opening Pandora's Box? Wish me luck. One last thought: Never leave your power unclaimed, and don't let others use it without your knowledge.

Unexpected Magic

This is the part where I get to show you the fun I had researching this book, and how many witches are doing amazing things that even MORE amazingly they were willing to share with the rest of us. This chapter had its genesis when I wrote up a questionnaire and sent it out to people, and additionally, asked most of the same questions in many different social media groups - and started lots of debates! Remember when I said this book was about starting conversations? It did, and it will continue to do so. Some of them, I imagine, will be quite serious; For right now though, these are the fun conversations that I had with witches of varying levels of experience leading up to the writing of the book, sharing their thoughts on technomagic, and ways they've used it themselves.

There are lots of ways to use common technology in uncommon ways, and there are people all over the Internet that are coming up with new ways to use those technologies as well as emerging ones - every single day. (Please note, I've listed names where I can**, and links for everyone I've mentioned that are content producers of some sort. Do yourself a favor, check them all out!) By no means do I mean this section to be a spell book, and I'm not by any stretch taking credit for any of these other than where I clearly identify it as my own. These are here as a way to make you look more closely at the tech around you, and see if there are new or unusual ways to use it that recommend themselves, or call to you, and act as a celebration of the resourcefulness and creativity of our witchy contemporaries. Ready to dive in?

I'll start with one of my own: As an artist that has worked in both digital and traditional forms for over twenty years I've frequently hidden sigils under something I've painted...whether it's a digital image, a watercolor…or a wall in my house. The idea occurred to me after reading an article on kitchen witchery and someone was making a pie - and they used a butter knife to lightly trace a sigil into the crust, right before adding the pumpkin pie filling to the crust to bake. The family would eat it - and none would be the wiser. Thanksgiving gatherings were stressful for this person, with so many family members of disparate ideologies gathered in one place, and it was a plea for peace. I loved the thought that others who normally disrespect, or disparage the idea of magic sitting there at the table all chowing down on a 'magic pumpkin pie', thinking how wonderful it was that it might work, with those eating having no clue what they ate!

At some point, I remembered that pie, and wondered if I could do the same with art. I did it, and was incredibly pleased with my results (and myself, lol). I thought I was being SO inventive incorporating spells into my art. Guess what? If you read the Virtual Spell Exemplar chapter, it becomes apparent that others have done it already! Another thing I've done that's quite effective is to create a sigil or fully type out a spell - and then SHRINK it down, embedding it inconspicuously into a picture. Most are probably totally unaware of what they're seeing unless they have specifically been told it's there. And like many other things, once you've seen it...you can't unseen it.

This ISN'T new. Regardless of what we are or aren't aware that we see, our minds see and register everything, even if we aren't consciously aware of it. Think about it: that's the whole premise behind subliminal advertising, right? Think of a goal an ad agency seeks to achieve - a message such as, 'buy more liquor', 'people will like you better if you buy (fill in the blank)' or they 'won't like you if you don't buy (fill in the blank).' The messages are hidden in plain sight, but not registered as 'seen' by most that view the ad. It's taking areas where a person may be vulnerable in terms of their self-esteem, and reinforcing it negatively with carefully chosen imagery. That's how traditional subliminal advertising has always worked. They knew it over thirty years ago, and it works just as well today. Subliminal advertising was always a hair's breadth away from being magic; Today, add the internet, and a sigil, image or spell hidden within, and you've got subliminal magic. I mention it because several people responded by questionnaire that they were using subliminals and sigils together, in memes and printed materials, specifically as subliminal magic. Whether the sigils are well-hidden, or in plain view, the people who view them register them, whether they are conscious of it or not. When I've used the technique of embedding sigils and spells, it's more often as a binding against someone doing hinky things, or as protection for someone who moved directly into harm's way. Beyond those uses, imagine creating a subliminal image like that, that's meant to be forwarded (Like to charge, repost to cast) or seen by many people, and suddenly your spell is doing double duty.

Another awesome way to use tech, you probably have right beside you, in your pocket or close at hand. Your cellphone! Yes, we've discussed ways to use your cellphone to cast emoji spells and other kinds of spells. Have you considered using your screensaver and home screen? My screensaver and home screen both contain sigils - one I created, and one I didn't - and I used them to turn my phone into a digital talisman. Your cellphone can be your ultimate techno-magic tool. Imagine: It's with me all the time, it's a continual visual reminder, and every time I unlock my phone, I'm recharging my intention! I learned this in a class I took with Rachel Capurso, who owns Aeolian Heart Astrology. Rachel has two groups on FB, a Patreon group, or you can visit her at aeolianheart.com. It's a great group of people, and there are awesome classes!

As I was prepping to release this book, I took part in an amazing magical experiment. Rachel and her husband Andrei Burke and Chaweon Koo of 'Witches and Wine' worked jointly to put together a ritual that utilized both ancient and cutting-edge magics, which I was thrilled to be part of. I felt I needed to mention it here because it's a good example of technomagic. Additionally, it's a great example of combining the old and the new to take advantage of both. The ancient part of the ritual was the astrological timing, which is one of the oldest and most traditional ways to use magic. At the exact moment of the Mercury Cazimi on June 10th, a spell/petition was inserted into the permanent metadata of an Ethereum blockchain, which is the cutting-edge part. For those who might not know, Ethereum is a cryptocurrency, and the 'blockchain' is like a digital transaction record that's written in a way that makes it very hard to cheat, hack, or steal it. The metadata into which the petition was inserted will now be a permanent part of that currency. The ritual itself has ended, but the magic will continue. We'll see where it takes us. (See the 'Helps' appendix for links.)

What follows are the comments and suggestions from some of the folks I interacted with, printed with their permission, and attributed in the ways they requested. I consider myself privileged that I got to speak or otherwise communicate with so many awesome folks as I continued to research, particularly as I realized delving into how people do magic can be especially personal. I asked lots of questions, and talked to practitioners of varying skills/knowledge. To those who filled out my questionnaire and sent it back, and to everyone who took the time to talk or chat: thank you! It is appreciated.

To start off, Spectregnosis@tumblr.com, has an awesome and informative blog called "Eternal and Eldritch" on Tumblr, with many awesome ideas for technomagic. One of the things I thought might be especially useful was his instructions to create a digital amulet. It was from his series on "30 Days of Super

Simple Spells" and was featured on Day 19. I made one for myself. Here are his instructions to create it, shared with his kind permission. The spell requires:

A blank USB flash drive

Internet access

A way to make the flash drive into a necklace (my note: when I searched, there are plenty that are set up to be on a keychain, or have a place you can hook it onto a lanyard or inside a purse or book bag. That's what I went with.)

1. Plug in the USB device.
2. Make three blank folders, titled as so:
   THE_SOURCE
   THE_PURP
   THE_FINALE
3. In the first folder, you will place images, texts, and anything you think will power the amulet. You'll want to use signs of power, sigils, galaxy pics, basically wherever you feel that the magical spark will come from, text files with things you find to be pure energy and mp3s with songs that pump you up are also encouraged.

The second folder will contain one thing, the program of the amulet. You will write out a .txt file and then switch the font over to wingdings, or something to make sure it's unreadable. (We want this to create a sigilized effect.)

The final folder will contain the finale. Here is where we place the trappings of the amulet. Like the first folder, you can add anything you want, just so long as it reflects the intention of the amulet: Certain gods in line with your intent, pictures, mp3s, poems, stories, and pdfs of books about the intent. For example, if your intent is love, a Jane Austen novel; for protection, a book on self-defense, or spells of protection; for courage, a biography of famous generals. There is no limit to this stuff. (My addition: make sure it means something to you! Everything magical works on symbolism, so for this to be most effective, the items you put in here need to be near and dear to your heart, or hold meaning to you or are significant for you.)

4. Place ALL the files / folders into a NEW folder you create called "FIAT_VOLUNTAS_MEA", which translates to "My will be done" in Latin. It ties the amulet together, to create a link between the folders, and power the amulet.

5. Decorate and carry the amulet with you.

NOTE: Guess what is inside a USB drive? A circuit board microchip. Guess what that is made of? Silicon. Guess what else is silicate material? Most of the planet, including quartz: that fun thing in your radio, watches, smart phones, computers, and hung around a lot of your witchy necks or carried in your pockets. The idea of charging a USB drive isn't so different from charging a crystal, except I feel that it's more hands-on, at least in the manner I've presented."

"Life is change. If you're not changed by the experiences

you have, there's something essential missing. I know my own approach to magic has changed quite radically in the last couple of years, and seems to be continuing in that vein. I use tech a fair amount in my magical work, including using social media interactions to fuel magical workings, or playing video games as part of a magical working, to using graphics programs with sigils."
Art Magic: Put together a collage - either online of photos from the internet (using Photoshop or some other graphics program) or from magazines, using glue/tape, and then taking a picture and sending it via email to a select group of friends - to charge and cast. Though the collage is activated by the viewer, the goal isn't to make it viral; it's to send it to a group of friends that you know will add their intent to it. To cast, the email you sent your friends schedule a time meaningful to you, and (whatever that time is) everyone sends it back to you at the same time (including you, sending it to yourself.) Sending casts it. You can also create a collage of sigils as well, and use the same methods just detailed. Taylor Ellwood (See the appendices for links to Taylor's blog, and a link to his books on Amazon.)

In a group on Reddit, user AnthropicPrinciple42 announced a project he created called "So Easy, It's Magick." It's a webpage that the author says is a technology/magick hybrid, and when you type in a spell, it will cause the program to attempt to execute it for you…in the spirit world. It gives you a place to type in your spell, and then you hit 'add'…as if it were a program. I tried it out, and took screenshots of it, which are in the Photo Gallery. †[15] There are two things that struck me about it; first, I'm fascinated by it, that I could

[15] See Photo Gallery, pictures 13 & 14.

put my spell into a text block, check a box and hit add, and a computer program would then decide whether or not to run my spell. Second, I want to know more about it. Will the spells actually work? Is it still my intent that powers it? The creator plans to come up with a way to create egregores using the same platform, and has heard of people using similar AI programming as a communications device from the spirit world to ours. I think it's still a work in progress, but if you want to try it out for yourself, there's a link in the Resources Appendix. (Also, a caution: the webpage has you interacting with a deified version of the Internet, which I didn't realize at first. See more about that in the closing chapter.)

My BOK, aka BOS is on Evernote. I use an app called Fluid to draw sigils. Black phone screens make great scrying mirrors. I write messages to my dead family members through Facebook posts. I set intentions with my backgrounds. I downloaded a digital compass to go with an experiment I'm working on, and I've participated in a zoom ritual. That's all I can think of at the moment." Victoria Pike

Another person told me of a simple thing; it's a wonderful way to do a spell in plain sight using a screensaver that your computer likely came with. You know the one where you can put text on the screen, that most people ignore? Here's a fun way to use it: Make your text and the background color the same. It's there, and you're the only one that knows. Or make the colors *almost* the same - your text is only a shade off from your background. That way, you can see it if you're close to it (so it reminds you that it's active) but the people around you likely won't notice it.

"Technology at times is an outward reflection of my inner journey as a witch and creative business owner. I first approached it like a grumpy old lady because it was intimidating and frustrating. Now, as I've gained experience and confidence, technology makes so much magic possible. Everything plays a part in running my witchy business, from electronic payment capabilities, zoom meetings to Photoshop for my digital works of art. I'm finally at peace with it and myself." Angelia Scott

Someone I spoke to in one of my groups uses "shufflemancy." What is shufflemancy? It is a form of divination that uses music, songs and melodies - even the titles of songs. Create a playlist on a cellphone; It should have at least twenty songs - but more like fifty is better. Bring up your playlist, and focus on the question you have. Hit shuffle, and listen to whatever song randomly comes up. Other options for shufflemancy are to decide a number of times to hit shuffle before listening to a song; putting a variety of songs and genres in your list...or just picking favorite songs...or using instrumental songs, and their energy to meditate your way to an answer. Take your pick!

Here's a unique approach to exacting revenge when necessary:

"Basically you can write one really good curse, and then just attach it to a spreadsheet and add names as you go. I have a different spreadsheet, where I have list of people who are using a servitor I made that releases old trapped negative emotions from everyone on the list. That servitor then sends those removed emotions to the people on that

curse list, and dumps them into those people. I've been tracking the results on around 200 people for a year - and it's worked outstandingly." Rich Kiska *(Additionally, Rich has experimented with a Healy bio-resonance device that uses micro-frequencies for removing emotional blockages as well as manifesting.)*

A lady I chatted with had an interesting take on emoji's that suddenly made them look less simple:

"I use emoji's like tarot cards, or any other form of divination. Look at each emoji. Really look at it. Somebody had an idea for it, to create it, you know? It went through a process; someone took it from an idea, to creation; and then others took it from programming, to using it. Whoever created each emoji had SOME purpose in mind for it, something from their own life or their own experience that gave it its' own individual meaning - which makes the emoji's more like divination. If you create an emoji spell with that in mind, and you contemplate each emoji before you use it, it makes them more useful, and your emoji's in the spell will resonate with each other better."

More than one person mentioned the use of Sound Frequency videos that are on You Tube, and several use the Schumann Frequency to help ground themselves. I've listened to them myself before, but didn't know how they were supposed to work.

Binaural beat audio will have one tone (or music) at a particular frequency going into one ear, and another tone (or music) at a different frequency going into your other ear. Your brain then interprets the two tones as a beat of its own, that is in-between the two original frequencies. As you listen, your brain activity matches that in-between frequency, and is helpful in training your mind to reach certain mental states.

The Schumann resonance frequency is 7.83 Hz, and it's in the Extremely Low Frequency (ELF) portion of the earth's electromagnetic field. Those who listen or meditate to it, or ground themselves with it claim many benefits, including better memory and ability to learn, less jet lag, resistance to mind control, improved tolerance to stress and the ability to live a more balanced life.

"I have always been fascinated and drawn to tech, because to me, tech is magic personified. Using symbols (1s and 0s), subtle forces are controlled and used to do amazing things, including controlling physical objects, physically guiding us via GPS, allowing immediate communication all over the world, etc. At 50 years old, I have seen first-hand how IT has caused the fastest, most profound economic and sociological change in human history...and it is only accelerating. It is opening the door to real trans humanism.

And as successful magic requires the practitioner to change to minimize potential negative results, mankind itself is being faced with the same challenge. Tech will either free us, or allow old patterns of oppression and bigotry to become enhanced. And I suspect, that just as models in our mind (spirits, angels, demons, gods) are also entities in their own right - another rabbit hole to explore! - I suspect that the archetypes will remanifest in new forms that are more aligned to the context of technology. We are seeing this now. Superheroes are the new expressions of the entities and archetypes of the past." James M.

On Tumblr, there was a long discussion about witches using QR (Quick Response) Codes as sigils, seeing them as useful because people see them everywhere - but mostly ignore them. In terms of magic, what some ignore, others will see as an opportunity. Several people mentioned them; apparently, they are used often in a game called Animal Crossings New Horizons for Nintendo Switch, while others are using them as spells. I played around with it a bit myself. I used a QR Code to create a link to take you to my webpage, which you will find with the rest of the links.

"My alchemical journal is on Google Drive, I have a Mercury symbol as my screensaver, and I have done rituals on my tech. A virtual altar is a great idea! I guess the fact that I like to do my work virtually makes me a bit more of a Tech Witch than I had ever thought before." Tara Plas

Subliminal Magic/Messaging was mentioned by one respondent, who said that she used subliminal magic, in some form or another for nearly everything she did. Most of us are aware of subliminals, as they've been used by advertisers/ad agencies since at least the 1950s. The premise is that a message (whether auditory or visual) is inserted into media below the "Absolute Threshold Level" or ATL of our conscious awareness. Famously, frames of "eat more popcorn" and "drink Coca-Cola" were inserted into a movie to see if it changed the behaviors of the viewers. While it's now in question whether the data was fudged,[16] the fact that subliminals are still used and we are exposed to it regularly is incontrovertible. One of the more famous examples is when Warner Brothers admitted to adding frames of a full-screen death mask in the movie "The Exorcist." I don't know about you, but having a VCR capable of frame-by-frame forwarding, my sister and I went in and looked for the frames of the death mask. They're there, and flash up during some of the scarier scenes to make the movie seem scarier.

> *"Want to manifest something? Write it on a sticky note, slap it on your computer monitor, then boot it up. Want to hotfoot someone? Nuke them in the microwave. Similarly, you can freeze a situation or slow it down in your fridge. Confuse someone by tumbling them in your dryer. Or, send your magick via text (Text "my new home is coming to me easily" to your coven mates.) The internet is the most magical tool yet, tbh…" Jessica Moore*

Electricity is a powerful energy source, and it's literally all around us. Two folks mentioned meditating with electrical currents. All the electronics around us have them, and we are electromagnetic beings. If we weren't, we wouldn't be conscious. I tried researching this a bit, because it's intriguing to me, and there's very little on it. (All I keep remembering is the movie "Contact", where her friends are telling Jodie Foster's character that others are laughing at her for listening to the electric of washing machines, and reminding her

---

[16] https://visme.co/blog/subliminal-messages/

that some call her the 'high priestess of the desert,' and laughing at her.) There's quite a bit on regular meditation, and feeling an 'electric' sensation - but not so much on meditating, say, on your television or light, and feeling the energy it either emits or uses. One thing I read that also intrigued me was that "Our thoughts are magnetic; our feelings are electric." Together, they make up our personal electromagnetic force. One person went further, and recommended that if you try it yourself, focus on the energy our electronics emit, and also use; focus on holding them, and letting them go. Once you can control their energy, you can use it. All I know for sure is that I'm intrigued. I'll be experimenting with this, for sure.

A good friend of mine who's been doing technomagic in some form or another for thirty years said:

> *"Energy is energy. It can travel through the air, or through a Cat 5 cable. It's all the same." He also noted "Especially in the traditional community, there will always be these debates. Can you use a wand you purchased or do you have to go out and make it yourself? Casting spells and doing rituals online were debated when there were BBS systems. I always say, "You do you!" Unless you are in a traditional coven it doesn't matter; If you join a more traditional coven, then you follow their rules."*

Multiple kitchen witches mentioned their freezers, their microwaves, their crockpots...and their KitchenAid mixers. Said one respondent, "A KitchenAid isn't a magical tool?? Try prying it out of my hands!"

Some folks questioned what counted as technology, because it quickly became apparent everyone has different places of drawing the line when it came to appliances. Some people felt if the appliance in question has electric going to it, its technology. Others said if it was connected to the internet or a computer, it then counted as technology. But what of where these technologies overlap? With the Internet of Things (IoT), and so many 'smart objects', it's not unusual to have a smart refrigerator or TV, and even our cars have computers in them. My thought on this is simple: if you can find a way to do magic with it...rock out with your bad self.

I'm going to close this chapter with the comments from a colleague, Thoth888, who has been practicing for quite some time, and I defer to him to have the final words:

> *"Tech? It's a natural, normal progression. As you learn new things, isn't it normal to integrate them into your daily activities? If you live in a technological world, wouldn't it be stranger for magick to not develop a technical bent? Nothing stays the same. We don't live in a vacuum. We live, we grow, we change. We are the sum of all our experiences in this life and the previous ones before this. If our magick didn't change as we learn and grow, I think it would cease to be relevant to our experience. "*

> *"Magick, like money, isn't intrinsically good or evil. It's dependent on the intentions of the person who wields it. Knowledge - gaining more of it, and learning to put the knowledge we gain to good use by our experiences and thus*

growing spiritually and mentally - is always preferable to operating blindly. If we give no thoughts to the ramifications of our actions, or unwittingly use magick, our results will likely be unexpected or outright fail, at best. In the worst case, our results could potentially be harmful or disastrous - leaving us to deal with the consequences of our actions."

"What's my favorite techmagic to do? Whenever I take pictures with my digital camera, particularly when I'm taking pictures someone asked me to take; friends, family and people I don't know, I always add my intentions for their best and highest good. That way, whenever they look at the pictures I took, it reactivates the magic I intended for them into the pixels."

"What would I recommend? We're at the forefront of great change, and with this change comes great opportunity. Learn. Soak in new knowledge, and assimilate it, incorporate it with what you already know. It's the best way to thank those who came before us; those on whose shoulders we stand. With every great increase in knowledge and power, there is usually uncertainty, and fear-mongering that accompanies it. Don't be caught up in it, if you can help it."

The Intersection of Games and Magic

"A useful way to work magic can be found in simulated online worlds such as World of Warcraft, Second Life, Everquest and Final Fantasy XI. In these games you interact in a simulated environment with other people from around the world, which offers a chance to work magick with those people."[17] *Taylor Ellwood*

When most people think of games, they don't usually think of magic. Some of the stereotypes people DO think of aren't nice (14 year-old nerds, unemployed 38 year-olds living in their mom's basements, or people who can't deal with reality, anyone?) but *most* of us gamers are normal folks who want to get together with their friends after work or whenever we can squeeze in a few hours for camaraderie, and to take out the aggressions from our work days on unsuspecting Orcs, opposing factions or anyone willing to fight you in PVP (Player vs. Player) mode. Every game has ways to use it magically, if you look for it.

SECOND LIFE

Second Life has a special place in my heart, which is one of the reasons why I'm putting it first! Second Life is a product of Linden Labs, and came to be in June 2003,

---

[17] Pop Culture Magick: A How to Guide to Practical Pop Culture Magic, by Taylor Ellwood, © 2018, 3rd edition, pg 117

and it's less a game than a destination, though as I mentioned earlier, there are plenty of games to play, and lots of people who are into the same things; finding them isn't hard. When I first realized/decided I was a witch, I was very much in the broom closet, and coming out of it was an incremental process. My first little, itty-bitty toe out of the broom closet door was in Second Life. For me, it was Goddess-sent. In my real life, I was just normal old me. In Second Life, I lived openly as a witch. It never ONCE occurred to me that my spells in-game could be real, or that the sacred space I created for myself was truly sacred, either. I knew it felt lovely, and I had fun. I met lots of people, who didn't question my choices, my identity as a witch or anything else. For me, the biggest draw was that it felt wonderful to be ***me***. Real. Not hiding in plain sight. I remember at the time, it irked me that I didn't feel like I could be myself in my own life. (Note to those of you reading this from your broom closet: Do what you gotta do, to go through your life and deal with the folks in it; Be the witch you're meant to be, even if it's in secret or only with a chosen few, and hopefully, some of the methods I mentioned throughout this book will be like a gift to you. I've been where you are now.)

I joined Second Life (SL) in 2005, while it was in its' relative infancy. In game, I had a digital altar in my digital house; I went to places that were witchy in nature, and while I didn't think I was practicing real magic in the game, it had its' own kind of magic that did something utterly amazing. There, in a digital world-between-worlds, I felt like a witch. That counted for much.

Fast forward another ten years, to 2015. I had inched out bit-by-bit from the broom closet, to the point where some of my friends knew, my family mostly didn't, and beyond that, I didn't worry who knew and who didn't. Real life took over my time for a digital one. I stopped hanging in Second Life, but I always looked at it fondly, and occasionally wondered if any of it was *real,* other than how it made me feel. I came to the conclusion that it helped me pave the road I went down in real life; maybe the magic involved wasn't only in my mind.

Fast forward five more years. At the beginning of 2020, I published my first book and after a couple of months, screwed up all my courage, and posted to my Facebook page about the book that was recently published. And just like that, all the sacred cows got kicked over, since friends, family and random folks I knew from elementary school to casual acquaintances who were friends with me on Facebook could see what I was. I was no longer in hiding. It was out there now that I was a witch, for anyone who cared enough to look. As I began working on my follow-up book, I started doing research for one of the stories that had a techno-witchy flavor to it. (And now we're all caught up, and off of memory lane.)

I realized several things pretty quickly once I began researching on technowitchcraft:

There weren't a lot of books on technomagic, technomancy, technowitches, or how they practice; when the topic came up in my groups, there was some angst, and debates, even on terminology. It made researching harder, all around. The places that were freer about technomagic were Tumblr, TikTok, Amino and Reddit, as well as YouTube. One book I **did** find invaluable was Taylor Ellwood's book, Pop Culture Magick; it's excellent, and even goes so far as to cover how to take a game's "lore," and create your own magical system around it.

Part of the reason there's not as much on technowitchery, Pop Culture Magic, or even technomancy as there is on other subjects is that those who practice them are either flying under the radar, keeping their practice to themselves, or actively avoiding scrutiny or ridicule, and finally,

I realized that a good bit of what I did in Second Life all those years ago was likely as valid as anything *else* magical I'd ever done. **That** thought intrigued me.

Despite already being in the middle of writing that follow-up book I mentioned, I found myself realizing that I was doing 'double-duty' as I began researching a book - this book - into existence. I rejoined Second Life when I began researching technomagic. And once I was back in SL, I began practicing magic in-world again…this time more as an experiment. It's been fun. It always was.

Before going farther, a disclosure. I ***have*** to say this, though I really, really don't want to.

Second Life isn't easy for some folks. I'm a techno-nerd, and there are more computers in my house than there are people and critters combined, by about 2 to 1. My husband is a techno-nerd too - which you could probably guess, based on the ratio of tech to living beings! I've been fortunate; I had access to computers since they came out, I've been playing Warcraft since 2012, and I've worked as a graphic designer and 2d/3d animator for over twenty years. I'm relatively comfortable with computers and tech. Second Life, for me, had a learning curve, but in no time, I was enjoying being a resident. I tend to forget that while I'm generally tech-savvy and comfortable with computers, *my* threshold of what constitutes 'hard' is a little jaded/skewed. As part of being thorough before recommending that everybody join Second Life, I decided to do a test. I enlisted the aid of my husband, who was, among other things, an IT guy in the US Air Force working a computer help desk, and has been playing World of Warcraft since it came out. I asked him to download Second Life, and create an avatar. He found it to have a steeper learning curve than he anticipated, and suggested that someone who is either not especially fond of tech, or may be 'computer-light' could find setting up an avatar and learning to walk, run and fly in Second Life either really annoying, frustrating, or daunting - possibly to the point of giving up. If you've got a good grasp on computers, have been a gamer for years, or you're really tenacious, you'll do fine in SL. If you're not, I'd still say give yourself a chance to try it and see for yourself - but SL won't (sadly) be for everyone.

In Second Life, I have two avatars, which are how a person's in-game persona is referred to. My original one that went back to 2005 is named Carissa Fitzpatrick, and my new one is Ellie Skydancer. Thanks go out to the awesome Linden Labs tech that helped me get my avatar back, and all her inventory, even after all that time.

The best way I connected with other witches in-game was to visit the stores that sell witchy items. There are two categories of witches (as far as I can tell) on Second Life: the real ones, and the ones that role play. Sometimes, the categories blur, but usually it comes down to whether they are witches in real life. Most of the ones that are actually practicing magic IN world are also practicing magic in their real lives. There are tools for the role players that tend to be flashy and have lots of particles (think wands that, depending, when you cast a spell will spout bubbles, butterflies or glitter) which take up memory, and truthfully, tend to get on people's nerves when you're in public places using them. Or 'teleport' devices that move you out of an area with swirls of smoke and light. Then, there are tools that are made mindfully by witches, to be used for real…in-game, to further your goals in your real life.

There are offering bowls that allow you to fill it with your choice of offering (salt, grain, or honey, for instance). There are candles that you can light and put out by touch: the colors can be changed for your correspondences, and better still, you can set a timer on your candle, and it burns down as the timer counts down. Additionally, you can type your spell into the script that runs the candle, so that as the programming 'burns' it, your spell activates.†[18] How many times in real life did you have to put out a candle sooner than you wanted to, because you have pets (or familiars) that could accidentally set your home on fire, and you needed to go to work, or bed? Or you ended up moving your candle to the sink in your bathroom or kitchen to allow it to burn out? (I've done that. Now I put my candles behind my fireplace screen, and my cats happily watch them burn.) There is virtual incense that you can watch the swirls of smoke rise. There are singing bowls tuned to the chakras that you can use in real life to balance or unblock them. Every tool you can have in real life, there is a virtual counterpart, made by a witch with intention, meant to work alongside your own intentions...like most of the physical tools you buy. If you're a 3d modeler or animator, you can create pretty much anything you like and upload it into the game. If you happen to be a programmer, you'll find that you can add spells into your scripts. With every new witch that joins, depending on the knowledge they possess, there are new ways to expand, and new tools that take the technology further.

In addition to virtual covens, there are virtual

---

[18] See Photo Gallery, picture 17

teaching areas where you can not only learn more about technowitchcraft, but also traditional witchcraft. Imagine, you sit down in a space, and interact with a teacher, to learn about magic as if you were physically there! It's wonderful how inclusive the communities are. They are a hodgepodge of witches, beliefs, and deities…and it works. I've noticed that the witches in-game are quite helpful, and are generous with the information they give - not only with magic, but tech issues as well. For example, while researching this book, I walked into one group's spaces without knowing a single one of them, and despite not knowing me, they not only had a lovely discussion with me about techno witchcraft, and showed me around their spaces, but were quite patient with me as I had issues connecting my headset to Discord, and ended up helping me set up a different viewer (Firestorm/Phoenix. I'd highly recommend Firestorm if you choose to try out SL.)

So, if you're a coven in Second Life, where do you go when you want to gather? While groups have public spaces, they also tend to have digital groves where they can work together. Having those spaces is actually quite important. It goes beyond the obvious shared experience, or camaraderie from the working, and runs headlong into reality. Imagine you want to do a spell in real life, which requires you to leave something at a crossroads, or you need to bury the remains of your spell, but don't want to bury it at your house. What would you normally do? You'd likely find a public park, or go for a hike in a local woods, or find a hopefully not-so-busy crossroads, and do what your spell requires. In real life, depending on where you live, that could present challenges. In Second Life, it flat out doesn't work. Why?

There are restrictions in Second Life on abandoning or opening objects in public places. Everyone is in-world to enjoy themselves, whatever form that takes, and like all societies, there are ground rules. One of these is that you don't create things and leave them for someone else to deal with, particularly if it didn't work! There are places referred to as 'sandboxes' that allow players to open objects, and test things they've created, sometimes with disastrous results, particularly if they weren't in a sandbox. It also allows creators to test objects that contain programming, but most places have very strict rules and permissions about what players are allowed to do, and where. If witches want to work together, they need a dedicated space that simultaneously allows everyone to be there, while keeping people who don't belong there out, that also allows them to set their permissions in such a way that people can do what their spells require. It sounds complicated, but really, it's not.

For instance, someone can come onto my land in Glimmermere; they're welcome to visit, and have a seat in one of the lawn chairs, even if I'm not in-world. (The SLURL in the appendices, if you click on it, will show you my house in SL; similar to Google maps, but for SL. Please note that *any* place I reference in SL will have a corresponding SLURL in the appendices.) Anyone who visits can enjoy the ambiance of my twinkle lights and the bonfire. They can't take anything, or leave anything, and in fact, if they try, it won't work, and any objects will be returned to them, as if they dropped it. My land in Glimmermere is on a server owned directly by Linden Labs, which every resident can take advantage of if they have a paid membership. Whether it's through Linden Labs, or not, all homes in SL have their own permissions that are much like a real-world homeowners' association. I can't use more memory than I'm allotted, so I may have to limit the size of the group I have in my space; I can block a specific person, but I can't block everyone, so that nosy guy that lives behind me in-world may decide to watch what I do, which serves as a reminder to remember to lower my shades while my avatar changes clothes! Having a space dedicated to group witchery is much easier.

If you're curious, and decide to try Second Life, there are lots of places that are witchy in nature. When I returned in-world with my new Ellie Skydancer avatar, I immediately set to looking up all the old places I used to go - and was saddened to find that many of them weren't there anymore, but new ones have sprung up. I began checking out new ones, and not only as research for TechnoWitch. I realized once I was back in-world, I wanted to be with other witches - but not in the timid way I had before. I wasn't a baby witch anymore. I'm no longer in the broom closet. I wanted someplace to belong, where people knew me. I wanted to find a place to be, and people to celebrate with. Don't be afraid to look groups up, talk to folks and join in! Look for people that have the same interests as you. That's what I did

Witches Fest Mall was there when I was first in world, and it still is, though in a reduced capacity. For a baby witch, it was a learning experience! Imagine going to a witchy mall, full of booths with witchy items (such as working tarot decks and runes), to home rentals in a witchy-themed area, to different looks for your avatar…from clothes, to skins, hair, all the way up to entirely new avatar bodies. They also had boxes of free stuff around, full of clothes and items a new virtual witch might need, which came in handy. There's an inn called the Artemis Tavern,†[19] where you can play games, sit around and chat, and even catch a class. At the center are spaces where they have rituals. FYI, this group is based in the UK, so most of the time **I'm** there, there aren't rituals going on, and it tends to be quiet, a bit like wandering the street at night. I've wandered around and bought stuff from the booths for hours. The SL group is sponsored by Children of Artemis (a RL group), and are part of the group that publishes Witchcraft and Wicca magazine in the UK. If only there was a real-life Witches' Fest Mall! They have a Facebook page, and there are several ways to contact them and interact with them in real life; links to all of these are in the appendices!

However, having said that, as I was in the process of finalizing this book, I learned there hasn't been much activity of late from Children of Artemis in Second Life. I ran into a lovely witch at the Artemis Tavern named Syra Sweetwater who asked me if I needed help. I explained I was trying to contact them. She told me that

---

[19] See Photo Gallery, picture 21

Children of Artemis in Second Life had lately been missing in action, and invited me to the spaces she runs with her best friend, who she met through the Children of Artemis group. It's called "Westwytch." The Westwytch grounds are beautiful; they have a party and gathering area, where the link at the back will take you, as well as ritual area.†[20] They have scripted 'dance' balls that control the motions of the participants, and visual effects during rituals. As each element is called in, lights swirl and dance. If you're looking for a fun group in Second Life, try this one!

Robin Wood's sim in Livingtree is my favorite. Remember I told you when I started in SL, I found a place I could be a witch while still in the broom closet? That was my favorite space to wander around. My avatar Carissa bought her first altar items in Robin's store, aptly called "The Broom Closet." †[21] She has an entire area dedicated to learning how to create items in Second Life, and dedicated quite a bit of her time in-world to that end, teaching many people over the years. Her enthusiasm for art, creativity and teaching benefitted all who came across her, including me. It took a while for me to connect the dots that "Robin Sojourner Wood" in Second Life was the same lady who made the Robin Wood Tarot Deck that I had hiding in my underwear drawer at home, who was the same lady that did the amazing artwork of the characters from Anne McCaffrey's 'Pern' books. I hoped to interview Robin

---

[20] See Photo Gallery, pictures 23-25

[21] See Photo Gallery, pictures 19 & 20

about technowitchery; When I contacted her in-world, she thought the idea for this book was a good one. She told me I could inbox her some questions, and she'd answer them when she was feeling better. Unfortunately for me, and for all of us, Robin passed away in April 2021. †[22] She and I never got to discuss technowitchery, and the world lost an amazing person.

"Can online games actually affect our reality and if so how do we choose to let them? As magicians, we believe we can shape reality, and simulated realities are in a sense a step toward that process. Does what happens in one reality affect another? I believe that this can happen and that even as people explore the idea of using the online network to form cybercovens and the like, online games are an extension of the technomancy we can use to shape the ways we work and understand magick."[23] *Taylor Ellwood*

WORLD OF WARCRAFT:

There are two versions of World of Warcraft: The commercial one that's either running the current content, or the "classic" content; or, the option of playing whatever expansion you choose on a private server. There's no right or wrong choice, only the choice

---

[22] See Photo Gallery, picture 18

[23] Pop Culture Magick: A How to Guide to Practical Pop Culture Magic, by Taylor Ellwood, © 2018, 3rd edition, pg 119

that's right for you.

Commercial: There are generally two ways I (personally) use the commercial version in terms of magic. The first is to do what would amount to real world equivalent of going for a walk. I go to places in world that I find beautiful, and feel peaceful to me, or go to places where I can meet up with others. As many people as there are populating Warcraft - particularly whenever new content is released - it helps to find places that have less people, which means generally staying away from areas where new characters begin, or the new content areas when they're released. One of my favorite places to go is Pandaria. It's beautiful, it's especially appropriate for ancestor remembrance, and usually isn't crowded, either.†[24] There isn't really a way to set up an altar; there are places where the quests/storylines have altars, which could be used. Particularly if you're someplace that is low on people, there is nothing that stops you from speaking your spell into trade chat, or if you're with others, being on Discord together.

Another way to use the commercial version of Warcraft magically are questlines†[25] that can double as a ritual - for instance, the Lunar Festival that honors ancestors, the Noblegarden festival celebrating Spring, the Midsummer festival of bonfires, Hallow's End, which celebrates Halloween, a Harvest festival that lines

---

[24] See Photo Gallery, picture 30

[25] See Photo Gallery, pictures 27 & 29

up with Thanksgiving, and the Winter Veil festival celebrating Yule.

Whenever I take part in these rituals that are part of questlines, I try to keep what the purpose is, and play along. For instance, the Hallow's End festival has a quest where you're supposed to throw several lit torches onto a Wicker Man setting it alight, as well as burning your cares. †[26] It serves double duty: As I'm throwing lit torches, I do it thoughtfully, and instead of just lobbing virtual torches at a virtual fire for points and gold, I'm taking the time to name my cares, and throw them on the pyre. Others, I've heard, will visualize a sigil on an enemy's back before they fight or kill them, using the pent-up emotions of the day to charge their sigil. If you can think of a way it works for you, guess what? It works.

Server: Did you love Lich King? Cataclysm? Pick the Warcraft expansion that was your favorite, and play it on your own, or invite your friends. Fair warning: It takes time and effort to get it up and running. There are helpful videos on YouTube on how to set it up, as well as where to find the files to run it, and both the files and commands to create objects in world. If you want an entire WORLD to yourself, you can have it. There is nobody to gank you, or troll you during a spell or ritual. You can create a building, put an altar in it, and do what you will. You can have a coven for a guild, and even create your own content.

---

[26] See Photo Gallery, picture 28

MINECRAFT:

Minecraft is another game that can be used to create a sacred space, though for me, personally, I don't like its blocky appearance. Some folks love it for their sacred spaces/magical workings. I know from talking to folks that there **are** real witches within Minecraft, simply because they're already witches and they play the game. It's much like Second Life, in that you have your own little bit of world, (or your own server) and you get to do what you like with it, and there's no real direction. I'm sure if you have your own home, your own space, your own altar, if you're into Minecraft, they can likely feel as sacred as my home in Second Life. If you like Minecraft, and it appeals to you, it can do what the other virtual spaces do, and you may find it easier, as well.

SKYRIM:

From what I can see, there would be plenty of opportunity to use Skyrim magically, and people are. (Disclaimer: as of this writing, I've been playing Skyrim for a sum total of three whole days.) Like Warcraft, you can play the game content - or not. Unlike Warcraft, you can have your own space and build it to suit yourself. The closest Warcraft comes to that is having a Garrison - but every level 3 Garrison looks exactly like everyone else's Level 3 Garrison. In Skyrim, with the Hearthfire add-on, you can build your own home/buildings.

Hearthfire can be purchased either through Steam, or in-game by traveling to the cities of Dawnstar, Falkreath, or Morthal, and speaking to the steward. Whatever city you prefer to have land near, if you give the steward 5000 gold, you get a deed for a plot of land and can do what you like with it. I could've just purchased the add-on through Steam, but I figured I'd learn more about the game, and the world if I have to physically get my character from the starting zone to any of those places…in one piece. I'm still working on getting to Morthal.

The biggest problem I can see isn't that you can't use Skyrim magically. You could worship any deity you currently do. Or, there are folks out there who see the gods and goddesses of the Elder Scrolls of Skyrim as real, and work with them…or they may work with the teachings of the Scrolls themselves. It was researching this that I learned even as there are arguments between witches about the merits of including tech in our workings, there are just as vehement arguments between those of us who work with deities exactly "what" comprises a deity. A question was asked of the r/pagan subreddit about those who are basing their practice off the cosmology and spirituality of the Elder Scrolls. The answer the poster received was telling: "We're generally not too keen on pop culture Paganism here. Our gods are real. This isn't a game. We don't get to make things up or use fictional gods."

This is another one of those arguments where the people arguing are missing the point, much like the group of people who were arguing about cauldrons and slow cookers. They don't realize the fight isn't between "gods" and "non-gods" or even "made up gods." The fight is between modern vs. traditional, and they haven't realized it yet. Rather than even try to engage with the poster as to why they didn't agree with it, in ego, they just smacked the topic down, chased the poster away, and closed his post for comments. Others asking similar questions on different reddits found similar results. One poster had a thoughtful answer: "The gods of Skyrim/Elder Scrolls didn't arise out of a vacuum. Many of them borrow heavily from Hindu belief." According to other posters, the actual teachings contained in the scrolls are mostly from Gnosticism. My favorite remark was from r/user MisfoldedProteins: "I started becoming more open-minded about all this when I realized that the ancient traditions we study *were* the pop culture of their day."

STARDEW VALLEY:

This is a farming sim game I learned about as I continued my research on gaming and magic. I ran across Selkie Tidechanger while trying to research more on magic in Minecraft, specifically.†[27] Selkie has a blog on Tumblr that is dedicated to the game, and its magical uses. She has references in her blog for correspondences by season for resources as they are available. There are also posts on spells you can cast, ways to use the villagers as poppets, and correspondences for which

---

[27] See Photo Gallery, picture 31

villagers can help with various spells. Her blog is informative, and several people have based their work in Minecraft on Selkie's work in developing similar magical systems. Links to Selkie's blog are in the "Helps" appendix.

Animal Crossings New Horizons for Nintendo Switch/Switch Lite:

This one kind of blew my mind as I was looking around for evidence of magic in games. Animal Crossings New Horizons is rated as appropriate for children as young as three; there is overwhelming evidence that the bulk of people playing the game are actually in their twenties and thirties. The premise of the game is to take the deserted island you end up on, and build on it. You can craft any variety of objects, from tools to furnishings, to buildings. You can spend your time there endlessly decorating, or just enjoying the atmosphere, or sharing links to your ultimate island paradise with your friends, so they can visit. There are people building whole witchy-themed islands, including temples and virtual altars. They trade (or sell) their creations using QR codes on places such as nooksisland.com, etsy.com, twitter, and reddit: r/horizondesigns.

OTHER GAMES:

We spent quite a bit of time discussing MMORPG games, but what of regular games? I have a puzzle game I have a magical use for; it's for my iPad, and it's called Astra Jigsaw. It's a nice puzzle game, but where it falls into the realm of magic is it allows you to create puzzles from your own photos.†[28] I'm currently searching for a house, and I have two spells active. I took a picture of the house from an online ad; I uploaded it to my puzzle program, which allows me to name it and comment on it. I named it by the address, and in the comments, I wrote a spell/intention, and then an emoji spell. Once I create the puzzle, then I play it...and I add my intent as I put the puzzle together. Additionally, I've also used it to add "oomph" to real life spells, where I take a picture of the traditional spell working at its' completion, turn the picture into a puzzle, and do it again to keep the spell active, and add more intent. It's a beautiful thing.

---

[28] See Photo Gallery, picture 32

Of Endings and Beginnings

Magic doesn't have to be stagnant. It doesn't always have to stay in the past. And what is magic? Magic is technology. What is technology? Technology doesn't just have to do with computers... Technology is any sort of thing that you do to leverage our human intelligence- our humanness - to have nature work with our will. That's what technology is. So let's expand our definition of technology-that magic is an ultimate technology.[29] *Chaweon Koo*

When I started out researching this subject, as I stated earlier, I wasn't researching it to write an introductory book on technowitchery. I was researching it to flesh out a story for the next book in the series I'm writing - a fiction book called "The Magic Around Us" - where a young woman who spends most of her real life in SL realizes she's a witch. She begins to find ways to use the tech she's known and loved all her life to create magic. Needless to say, it doesn't work the way she planned, if in fact, she was planning at all. She knew she wanted something different, and technowitchery was certainly that. The only problem is that the woman never counted on it being real, or touching her life in the real world, or worse still, the lives of others in the real world...and drama ensues.

That was the plan. I jumped into the topic of

[29] https://www.youtube.com/watch?v=qcoJrOzNBSU

technowitchery for the research, and stayed because I not only recognized I was already using some of it, but also saw potential. When I began my research, the only things I knew for certain that fell under the category of "technowitchcraft" was a general knowledge of emoji spells, though I'd never cast one, and a class I'd taken through Aeolian Heart Astronomy called 'CASHFLOW: Mercury in Gemini' that (aside from being excellent!) meant I had a cellphone that I had turned into a digital talisman as part of the class. That was it.

After doing research, I've had my eyes opened. I didn't realize how much technowitchery surrounds us in general, with most of us not realizing what we see or interact with. That includes me, also. It reminds me of one of those drawings that plays on line and negative space. Every time you see it, you see a deer in a meadow - until someone shows you the picture of the woman hidden in plain sight, also known as an ambiguous image. Or Verbeek's duck that turns into a tree when you turn it upside down.†[30] These topics are the same way. I see it now. I see technowitchcraft and magic all around me, and likely you will now, too, if you didn't already. Would you rather be a participant in using the information that flows around you, or be tossed and turned at the whims of others? I know what I chose! Much of what we see on social media are, in fact, low-grade spells, even if we don't always realize that's what they are. It reminds me of a meme I saw: "Me wondering

---

[30] See Photo Gallery, picture 33

if I should tell my devout Christian friend that her vision board, affirmations and candles is about three steps from low-key witchcraft..."†[31] It may make you laugh, but think about it. If a witch, who is familiar with magic can use technowitchery and not realize that's what it is, how many 'muggles' or 'normies' post things unwittingly, with emotion if not intent - not believing in magic at all, and as a result, reap unexpected consequences in their lives, never understanding why? That's not even counting topics like "Memetic Warfare."

In my opinion, such as it is, if there is such a thing as 'evil' or 'dark' technomagic, it would be Memetic Warfare. It's like subliminal advertising, but on steroids. Subliminal advertising has ALWAYS been just a hop, skip and a jump away from magic. Ad agencies spend insane amounts of money, hiding things in plain sight, attempting to sway your decision on a purchase, or even your opinions. The difference is at least when you're seeing most subliminal advertising, you're aware that's what you're seeing. That doesn't mean you won't be susceptible to it. Memes? Not so much. People will see them, and laugh (or get angry) and pass them along, not realizing their power. Imagine how many memes have magic interwoven into them, but now - imagine how many of these memes have been engineered specifically - on purpose! - to toy with our beliefs and emotions, who we support politically, what causes we support. It's somewhere in the no-man's land between Chaos Magick

---

[31] See Photo Gallery, picture 34

and PsyOps. There are groups of people that - as their job - spend their days creating memes that seek any means to either change your mind or influence your opinion on hot-button topics, whether they happen to call it magic or not. Beyond that, there are companies that are using 'bots' and artificial intelligence to target these memes like a laser beam at the people who are most likely to be influenced by them. Giant corporations hire these companies; smaller companies do, too. Public relations firms use memes as a way to gauge how to help their clients, and if their strategies are working. Pick your social platform - it doesn't matter which. It doesn't matter your political leanings, or your religious beliefs or lack thereof, or if you do technomagic or not. It's there; all of it. If you are aware of the wild magic swirling around you, it either loses its effect…or you can use it for yourself and harness it, like electricity, to your advantage. That, or you can choose to ignore it. Would you rather use that energy…or BE used? I would rather use its' energy, and turn it into something constructive rather than be at its' mercy. Look these topics up for yourself, think about the things we've discussed since the beginning of the book, and form your own opinions, keeping this in mind: the things we ignore often increase and change, and are rarely ever in our favor.

"He that will not apply new remedies must expect new evils; for time is the greatest innovator."[32] *Sir Francis Bacon, Sr.*

---

[32] https://www.brainyquote.com/quotes/francis_bacon_130602

The topics that open up as we discuss technowitchery only ***start*** here. That's why at the beginning of this book, I said this was a conversation. The topics surrounding technowitchery go where we decide to take it, and as we change, it will change as well. Using the Internet and virtual spaces aren't bad, unless we make it that way. Mark Pesce said it best in his interview with WIRED magazine:

> *If computers are viewed as evil and dehumanizing, then we made them that way. I think computers can be as sacred as we are, because they can embody our communication with each other and with the entities - the divine parts of ourselves - that we invoke in that space."*[33]

Love it or hate it, tech has become part of witchcraft, and it's the normal flow of things - and I daresay it always has been, as witches co-opted what passed as the 'tech' of the current day. Witches are an inventive lot, who will find ways to make magic or be the magic even in the worst of times - which is why we scare folks that prefer people to fit into their assigned roles instead of forging a new path. Witches are trailblazers by nature, and improvisers out of necessity. You use all the tools at your disposal, and you adapt.

In closing, I read something that challenged me one evening on Reddit. I ran across a group of technoshamans. The questions they were asking one

---

[33] https://www.wired.com/1995/07/technopagans/

another are well, well beyond whether or not an emoji spell, or a meme spell, or turning your phone into a talisman is a valid, effective use of magic. It's miles beyond that.

Is the Internet as we know it now, where so many people are connected to it, and so much magic is being used (knowingly or not) within the fire of electric that powers it...actually a deity? Has humanity invested enough time and attention corporately to the Internet that it has gained a consciousness or awareness along the lines of an egregore? When I started writing this book, my first iteration of this chapter said:

"If people are asking questions like these in a public forum, that tells me that somewhere there's already people who are working with or worshipping something that they feel when they turn on their computer, laptop or phone. They feel its' power as soon as they turn on their device, and all that lovely, gooey information the Internet contains comes rushing at them instantly."

And then just like *that,* (snaps fingers) I had to change what I'd written. One of the examples I included in Chapter Five proved this out, spectacularly. Remember the webpage I talked about, called "So Easy It's Magick"? I wasn't kidding, I thought it was a cool idea, and I'm curious by nature…so I started doing simple research on XaTuring.Runtime. I figured it was: a) some kind of programming language, since it ended with '.Runtime;' and it has me putting my spells into the webpage like I'm coding something, and b) I know who

Alan Turing is; he's the father of modern computing. He literally invented the idea for the computer in 1936, and AI in 1950. What did I find when I looked it up? It wasn't what I was expecting, for sure.

A fellow named Don Webb and some of his compatriots did a working in March of 1993 to invoke the spirit of Alan Turing, with the intent of making him the "God of Cyberspace."[34] Fast forward to 2003, and someone named John J. Coughlin got really fed up dealing with viruses where he works in IT, and wanted to create a working that would get back at those who create them…and found XaTuring, a God-form already in progress, which he decided to use.[35] When you type your spell into "It's So Easy, It's Magick", it's XaTuring that you're beseeching for aid in obtaining your spell's successful outcome. There are many out there who feel that if Turing isn't the God of the Internet, then Thoth/Mercury/Hermes is a likely character/suspect.

Creating a Deity out of the Internet isn't so far removed from Chaos Magick or Pop Culture Magic. If you can take a character in a game, a movie or a comic book, or a cultural pop icon like Marilyn Monroe, and create a whole system of worship for any of them, then why not Turing or Mercury as representative of the Internet? Bystanders may not only be working with magic they don't understand or realize, but also may be working with a deity and not realize ***that,*** either.

---

[34] https://www.waningmoon.com/xaturing/genesis.html
[35] https://www.waningmoon.com/xaturing/intro.html

As this first iteration of TechnoWitch rapidly comes to a close, I offer this: this topic is only in its' infancy. It's left to you to decide ***how*** to use what you've learned, or even IF. There will always be purists, who will prefer traditional means and methods. Each witch to their own. All paths are valid. I hope those traditionalists who have read this far, even if they disagree with every last thing I've written, keep an open mind on the topic. They are surrounded by technology even if they choose not to use it magically. Better if those traditionalists who dislike technowitchery at least understand what they see and the possibilities of how it can be used (both for and against them) instead of *not* knowing. Otherwise, we are like that less-experienced witch who finds we are dabbling in things not understood, and reaping whatever the consequences may be.

Like it or not, we're continuously surrounded by forms of magic that may leave us with more questions than answers. Right now, magic in cyberspace is the virtual equivalent of the Wild West - and everybody's got a gun, trying to eke out a space and resources. As in the rest of life as we know it, there are winners, and there are losers, and sometimes it's what we know, and how we use what we know that makes the difference which category we fall into.

The question begs to be asked: How much further will technowitchery go?

Have we birthed a new way of using magic, or have we birthed a deity of the future? Has the internet given new life to older deities? There are any number of ancient deities that presided over wisdom, knowledge, and writing that are quite at home in the digital landscape. What form will the future magic we make actually take?

Stay tuned! We're creating our own answers!

(Join in the conversation by emailing me at technowitch.book@gmail.com, or PM me on Reddit @WitchyHeretic444

Appendix 1: Warcraft, Magic and living in the digital world – A conversation

This book was literally meant to start a conversation. We've started the conversation right here, ourselves. It starts off with games, but moves on to magic in general and how magic in the digital world works. I recorded it initially so I wouldn't forget anything we discussed, and I wasn't kidding in the dedication. My husband, my sister and I had some VERY interesting conversations about this material, and I adore bouncing ideas off of them. They're cool that way. Enjoy!

Ellie: What are some ways you can use magic in Warcraft?

Schoen: If you use the GM commands in the server version, you can create an altar, spawn up a fire pit, spawn up something to sit on. There are commands to be able to create different items. There's a catalog of item numbers, and you'd have to research what you want to use or create. That being said, you can create your own armor or outfits or items with various properties. (Lich King 3.3.5) You can create your own items for your own place, in your own game. You can modify items somewhat, but you can't totally customize them. But the fact that you're making an item yourself, in game, for your character – for your avatar - there's room for magic in that. You're crafting something for yourself in game.

Ellie: Not unlike a person creating, say, their own tarot deck or your own wand in al life.

Schoen: Yeah. Even if you're running the retail version of the game, there's places – there's empty houses in the game; you have to find them, of course. Some of them don't have any furniture in them, so there's nowhere to sit. The downfall with either the commercial version or Classic is that you don't have admin rights, so you can't create stuff. You can just pick an empty spot, and when you're in game, it's "your room in the inn." There's altars that exist in game, there's so many places you can go. You can type in your spell in the chat. Unless you have someone close by you, and they're watching the text, they're not going to know what you're saying. I prefer running LichKing as a server, because I like being able to be "god" in my game, to an extent. It takes some of the fun out of playing if you run in GM mode, or "god mode" all the time, and you decide to make your Lvl 17 toon into a Lvl 85 without earning it.

Ellie: That's one of those things I find really interesting about having your own version of the game. In theory, you could have... If you have a group of people that just wanted to use the game to have as virtual space to do workings...

Schoen: It's like having your own guild, in game. You get all your toons together, at an inn, or at a house or wherever, and you can say whatever you want. Or better still, if everybody's on Discord, voice chat...Now,

Second Life... You can do more there. Obviously, you set up your space. You buy your furniture, your tools, or make them. You can have gatherings of people, as many as you want. You can have sex even in the game, using the furniture (he's referring to the blue and pink colored poseballs.) Sacred spaces...you can create anywhere. You can find ways to do whatever you want, in whichever game you want.

Ellie: Sacred space, even if you're not in a game situation, you create it based on whatever you've got in front of you. Like, you could have a folder in your google drive, that you can access from any of your devices, right...? And have things that basically, to your mind "spell" an altar, to you. You could have pictures of your real one, or pictures of the tools that you use, or even would like to have. Or pictures of something that wouldn't seem magical, but to you, symbolically, it means something to you, and for you, it creates sacred space. There are altar apps. You could create a webpage. You could create a secret page as a sacred space in Pinterest! (Laughs) What if everyone creating pages on Pinterest are creating altars unaware that they are doing just that? In theory, that day we were getting the car fixed, and everybody had their faces in their phones. I could've been in my own sacred space; I was writing, but I could've been. (laughs) What's to say that when you're on your phone, all by yourself, that even something as simple as scrolling through Facebook – whose algorithms are bringing you what you want to see – does it create your own 'space'? It's you, interacting with your phone, which is another tool.

Some people aren't probably going to look at it that way, but...in theory, it kind of is. You post things you care about to your wall, other people post things, and FB goes looking for things it thinks you care about. My biggest thing, and the main reason, ultimately, why I thought about writing the book is because everybody has digital devices...all around them, right? Even kids have them. There are an awful lot of things out there, that are, by definition...either magic, or could be USED as magic and people don't realize that when they're doing something. I mean, you know what you're doing if you're using a tarot app, an astrology app or Book of Shadows app or some other magicky app, because you downloaded it to your phone on purpose. Great. But when you're on Facebook? And there's all these versions of emoji spells...

Schoen: Okay, so that's something to think about. So if you're on Facebook, right? Somebody puts a post up, and it's going to create an emotion. In theory, aren't all posts a kind of spell? You share something you like... You see posts, and you give them a heart...or a thumbs up...or a sad face, or an angry face. Aren't you adding your emotion to those? Everybody puts posts out there that bring out emotions, and any post becomes 'like to charge, share to cast". That's virtual magic, because that's emotion, and intent. Emotions are energy, so that's changing your energy, or sending your energy out there, if you think about that.

Ellie: Picking up your device, and then going into Facebook has more or less created a room, or a space...you're mentally IN that space, and then you're seeing things on social media...whatever the social media...that causes you to interact with it. Usually, they way they put it in specific magic memes, there will be whatever the content is, and then at the bottom, it clearly says "Like to charge, Share to cast." Think about how many Christian memes there are.

(Missy joins conversation)

Schoen: I have this friend. When you go to his page in Facebook, all of his posts are inspirational thoughts. That's all he has up there. There's this other friend of mine, who supports a particular political candidate, and everything on his page is all political, nasty shit, meant to be mean, meant to be negative toward the opposing party...and all his friends who think alike are all liking and sharing it. They're spreading their energy out there. Every time you like and you share, you're putting your energy out there. Every time. And their energy all feeds on each other.

Ellie: And that's the thing. People don't realize they're actually creating something – magic? Whether you count it an actual spell or not, they're still creating something. It's like, weather in cyberspace. Get a big enough group of people together putting their energy into a thought, it's like a digital storm.

Schoen: A cyberstorm! (all laugh)

Ellie: Exactly! Some of those people would never look at it that way, particularly the Christian ones, but what if some of them were riling folks up on purpose, knowing what would happen? Or maybe they didn't know what they were doing, but they liked the results they got. There are witches all across the political spectrum. When you think about memetic warfare, you're creating a meme that acts like a spell...that specifically targets a group of people.

Schoen: They're using social media – we already know this – using memetic warfare to get their points across. They specifically targeted those that were undecided, or were fearful to influence their vote. That's why that research company did all that. They found out 'this is the weakest link' and then they exploit the weak link.

Ellie: But if you consider the fact that every single thing you post can be, and almost might be a spell – when you think about memetic warfare...that's dark, taking it someplace dark. You may be cursing people, and you're throwing around bad intent, and you're putting your energy into THAT, if that makes sense. I don't really want to call it "evil" per se...but it isn't friendly, for sure.

Missy: It's bad intent.

Ellie: There are people that are using things in really bad ways. The average Joe might be using some of this, and not know what they're really doing. And they wonder why...

Missy: It's giving a child a gun.

Ellie: Yeah, kind of. That's why I wanted to write TechnoWitch. Hopefully, it can help people understand what they're looking at when they see it.

Missy: Until you tell a child to 'stay away from the

stove, it's hot', they're likely to get burnt.

Ellie: There are people out there that are wondering why is my life the way it is?

Schoen: It's what you make it.

Ellie: Because that's what you did to it. You didn't realize that's what you were doing to it, but you did.

Missy: Your reality is what you've spoken into existence.

Schoen: That's a good point. When I was stationed at Quantico in 1986-87, it was my worst. If something went wrong, it happened to me. If somebody forgot something…it was me. (laughter) Forgot a uniform item? It was me. If anything bad happened, it was me. I was "that guy." And I was like, "What the hell…? Why is this shit happening to me?" So finally, I'm like "Okay, how am I going to fix this?" I consciously decided to have extra of everything, to always be prepared.

Ellie: (laughing) Back to the Boy Scouts!

Schoen: And how does this relate to the virtual? It was conscious intent. Whether it's physical or cyber, it still comes down to intent.

Ellie: And that's the thing. Until you know… I mean, shit you don't know can hurt you. I was always told in magic "Don't do things you don't understand what you're doing." Don't walk blindly into something.

Missy: Remember what I told you about when Pop was on hospice? I was making prayer bracelets, and every knot was a prayer. I was doing knot magic, and I didn't even realize that's what I was doing. Some things, you may have an innate talent for, but unless you know what you're doing with it…you can be dangerous.

Ellie: You could be doing 'bad things.' You could be screwing yourself up.

Missy: You might not even realize you're working magic.

Ellie: Magic, whether you realize you doing it or not, is still magic.

Schoen: Okay, here you go. So you create a Facebook page. A private Facebook page...you don't invite anybody, or add any friends. It's your online space. Add whatever pictures you want to it, and you could get to it from any device.

Ellie: Here's a question, though. If you can create a Facebook page…that is for all intents and purposes, your altar…and invite nobody to it… What TRULY is the difference between your 'wall' and a virtual sacred space on Facebook, even if all your friends CAN see it?

Missy: Nothing.

Ellie: Exactly. It's semantics. Mark Zuckerberg could've called it an altar…but instead, he called it a wall. So basically, in cyberspace, every person on Facebook might as well have their own altar…even if they aren't calling it that.

Missy: The place where it gets unusual is paying attention to how many people you get on your page, and paying attention to who on there who hit like…or love…or hate - because that adds power. On some level, people know that.

Schoen: You can add pictures…music…videos you like…and it's all right there.

Ellie: The things people put on their Facebook page

are the things they celebrate; things that they're passionate about, that they care about; If you go to a Christian's page, it's going to be full of all kinds of memes that celebrate their faith.

Missy: That's why you join groups that have things in common with you.

Ellie: What's the difference on Facebook between some groups and a virtual coven, even if they aren't witches? (laughter) There ISN'T. What's a coven but a group of people who have shared interests, that may or may not work together, and have a similar belief structure?

Missy: Either way, they're still working together. There's a group I'm thinking of. When they were having the elections, and there were witches all over doing bindings?

Ellie: The Bind Trump group. *(NOTE: I've tried very hard to keep politics OUT of this discussion, so in all fairness, I feel like I must also state that there is also a Witches for Trump group on Facebook as well.)*

Missy: That's a lot of people with common intent… That's a whole lotta intent going out into the Universe.

Ellie: And that's why they're still working together. There are still people in DC that shouldn't have the jobs they currently hold. It's my opinion, but I think Trump himself uses magic. Here's the million-dollar question: Is he aware of it? Does he cultivate it? Is he advised by others in such a way that it results in magic?

Missy: Think about Twitter! How many people did he reach with Twitter, and inflame EVERYBODY?

Whether you were for him or against him, everybody watched his Twitter feed, if only to see what he'd say next. He made an impression. It was magic.

Schoen: On his heaviest day, he did like, 140-160 tweets…in one day. *(NOTE: it was 200 in one day, on 6 June 2020)* [36]

Missy: Isn't there some guy on his staff that is into the occult…?

Ellie: You mean Steve Bannon. He was one of the President's advisors. Yeah, he's into some occult stuff, from what I've read. [37]Either way, people are allowing stuff they might not fully grasp into their spaces; if you look at your wall on Facebook as being a virtual sacred space…you're letting people into that.

Missy: If you're a person who, on the outside is portraying yourself as wholly Christian…and then, you're actually doing some pretty dark shit on the down low…that makes you reprehensible.

Ellie: It does. And it makes you a danger to other people, I think. At least be honest with people.

Missy: If we're practicing magic, at least we're being open about it.

Ellie: There's a lot of people out there who aren't being open about it. *(NOTE: I wasn't talking about closeted witches, here. I'm talking about people who are being hypocritical in their beliefs.)*

---

[36] https://www.insider.com/trump-breaks-record-most-tweets-in-a-single-day-2020-6

[37] https://www.vanityfair.com/news/2017/07/the-strange-origins-of-steve-bannons-nationalist-fantasia

(pause in discussion)

Ellie: I'm trying to think of a way to say what I'm thinking. There are people who are good at everything. What if you are really good at getting likes and clicks and views...getting attention...

Schoen: You're referring to influencers.

Ellie: Exactly! Do those influencers realize that they might be doing magic? Some of them might.

Schoen: Maybe some of them do, but unless you believe in magic...

Ellie: Whether or not you believe in magic, it still works.

Missy: That's what I was talking about with the knot magic. I was doing traditional magic, and not even realizing it. It's been around for centuries! I just instinctively did it, without knowing what it was. And that's the thing. When you have people who are performing magic... and don't know they are...particularly if they have a bias against magic, it's still...

Ellie: It's magic. It's all intent. They just don't want to call it that. And that knot magic you talked about in your bracelet – it's not that different from some of the virtual spells. Every time you see the knots on the bracelet, it brings to mind the prayers you prayed – recharging them; kind of like a document or email that has your spell written on it that you don't open, and every time you see it on your desktop or in your email queue, it recharges it. An even older version of that knot magic is to tie your worries into a string or twine...and then burn it.

Missy: The more you look around you and see, the more cognizant you become. Media's a part of all this.

Ellie: What media adds to it? Media adds intent, as well. Every last news channel, for instance, regardless of the network – they all have their own spin on things. Every broadcast from a major network all the way to a YouTube channel – they're all adding their own intent to their message. And they're gathering people to themselves that think just like them. And they're broadcasting out what they believe, whether they're talking about creating a Skoolie *(a converted school bus that allows you to live off grid),* or discussing politics.

Missy: But that's what it comes down to; whether you realize what you are doing is magical. Because it's like...okay, if I know what I'm doing is magical, then my intent becomes paramount. Supercharged. Because I know what that magic is doing, and I'm doing it on purpose. It's with your intent.

Ellie: That's what makes me worry. Influencers that are actually feeding intent, and pumping intent out all the time, trying to get somebody, anybody to look. And some of the anchors on the networks...they're influencers, as well. Some of them are actual - very good - journalists...and others of them are nothing more than presenters or commentators whose sole purpose in life is to rabble rouse their viewers for ratings. Networks provide a platform to influence millions of people. A commentator can say something extremely provocative one evening, and in hours (sometimes considerably less) their statements will be turned into a meme, and weaponized. Suddenly, you have two groups of people sharing it...The group that agrees with whatever it says, and the group that's outraged by it. Both are throwing emotion into their postings and responses...along with intent, and raw power. Regardless of whether you believe in magic or don't, or agree with what the meme is about or not...both groups of people are sharing it, and giving it power.

Missy: That's what I meant when I talked about the President's tweets.

Ellie: You know, that might be a good way to tell if something's been weaponized. If EVERYBODY's talking about it. Good or bad, agree or not...

Missy: Intent is intent, and if everybody's talking about something, the intent behind the meme is amplified!

Ellie: After doing all the research I have, the one thing I've done is changed how I approach social media. I'm not on it much anymore, and I'm careful what I post. And when I *am* on social media, many of the 'agent provacateurs' I was friends with, I'm still friends with, but I've snoozed them permanently. I don't see their posts. I don't want to. Other than that, it engages me. I realized I was being USED, and I refuse to be used like that. I'm not going to "feed the bear" anymore. Instead, I'm writing TechnoWitch, and the bear can go to hell.

I'm imagining if you've read all this way back here to the Appendices, it means you're interested enough in the subject that you're reading our conversation. At least, I hope that's the reason. LOLOL.

The next book I write on this subject is going to be the conversations I have with YOU. What your thoughts are; how you're using tech with magic. What's good about it, and what's bad. What worries you about it and what we should embrace with open arms. I'd also love to know the cool things you're doing with it!

To that aim, here's an email address you can reach me at:

technowitch.book@gmail.com
and on Reddit: WitchyHeretic444
And on Tumblr: Mysticalarbitertyrant

You know what to do. Let's keep the conversation going.

BIBLIOGRAPHY:

Memetic Warfare:

Occult Memetics: Reality Manipulation, by Tarl Warwick, 2016
https://medium.com/tryangle-magazine/meme-magic-is-real-you-guys-16a497fc45b3
https://www.vice.com/en/article/xyvwdk/meme-warfare
https://www.aspistrategist.org.au/how-memes-are-becoming-the-new-frontier-of-information-warfare/
https://apps.dtic.mil/dtic/tr/fulltext/u2/a507172.pdf
https://densitydesign.github.io/teaching-dd15/course-results/es03/group03/
https://www.wired.com/1995/07/technopagans/

The Misadventures of Pepe and Kek:

https://www.splcenter.org/hatewatch/2017/05/08/what-kek-explaining-alt-right-deity-behind-their-meme-magic
https://theconversation.com/how-an-ancient-egyptian-god-spurred-the-rise-of-trump-72598
https://www.wired.com/story/witchblr-kek-online-occultism/

Subliminals:

Freud on Madison Avenue: Motivation Research and Subliminal Advertising in America, by Lawrence R. Samuel, c 2010 University of Pennsylvania Press
Dark Psychology Manipulation: Highly Effective Techniques for Influencing People Using Mind Control, Persuasion, NLP and Deception - The Subtle Art of Brainwashing, by Robert G. Peterson, © 2019
https://visme.co/blog/subliminal-messages/

Shufflemancy:

https://the-daily-diviner.tumblr.com/post/125219227364/how-to-do-a-shufflemancy-reading
https://aminoapps.com/c/pagans-witches/page/blog/a-beginners-guide-to-shufflemancy/YjRQ_EX8fbu1kRN1Y5Xzo3pN7WQBYbZz8PQ

QR codes:

https://www.victoria-westerman.com/blog/qrcodex
https://medium.com/@akramhelil/something-we-kinda-see-every-day-the-magical-qr-code-4f20bed179fa
http://www.monamagick.com/media/2014/06/qr-sigils/

QR codes specific to Animal Crossing game:

https://www.lifewire.com/use-qr-codes-in-animal-crossing-new-horizons-5095510
https://br.pinterest.com/sigils/interests-ac/
https://gamerant.com/animal-crossing-new-horizons-mostly-played-adults/
https://www.youtube.com/watch?v=3oQlotxFj4M

XaTuring/XaTuring.runtime/It's So Easy, It's Magick":

https://www.waningmoon.com/xaturing/
https://www.diva-portal.org/smash/get/diva2:1481545/FULLTEXT01.pdf
Https://itssoeasyitsmagic.com
https://forum.becomealivinggod.com/t/xaturing-working-with-the-god-of-the-internet-digital-sphere/5696

Alan Turing:

https://blogs.scientificamerican.com/guest-blog/how-alan-turing-invented-the-computer-age/

Meditating on Electricity:

https://meditationsecrets.org/electrical-meditation/
https://powerthoughtsmeditationclub.com/hey-you-yes-you-you-are-electric/
https://www.successtoolchest.com/blog/Effective-Communication-skills-The-Simple-Science-of-Feelings-and-Emotions

Pop Culture Magick:

Pop Culture Magick: A How to Guide to Practical Pop Culture Magic, by Taylor Ellwood, © 2018, 3rd edition
https://www.vice.com/en/article/d73q3x/the-pop-culture-pagans-who-draw-power-from-tumblr

Memes and Meme Magic (general):

https://heleneinbetween.com/2014/03/my-face-is-viral-meme.html
https://www.huffpost.com/entry/beware-of-hidden-messages-in-memesthink-before-you_b_5a12fee7e4b05ec0ae8443eb
https://www.heyalma.com/why-is-everybody-suddenly-sharing-these-alt-right-memes/
https://derp-cat.fandom.com/wiki/Meme_Gods

I found these helpful:

When, Why…If? By Robin Wood, © 1996 (ethics book)
Existential Kink: Unmask Your Shadow, and Embrace Your Power, by Carolyn Elliott, © 2020
https://www.iol.co.za/saturday-star/why-we-hate-new-stuff-2051383
https://www.business2community.com/leadership/how-leaders-use-symbols-to-influence-others-02096894
https://templeofanpuandaset.wixsite.com/templeofanpuandaset/online-shrine

Helps:

Aeolian Heart:

https://aeolianheart.com/

Link to CASHFLOW class that was referenced:
https://www.youtube.com/watch?v=ZBCk1vI_19I

Carolyn Elliott:

https://badwitch.es/

Chaweon Koo:

https://www.patheos.com/blogs/witchesandwine/author/ckoo/

How the Cryptocurrency Ritual was done:

https://www.youtube.com/watch?v=qcoJrOzNBSU

Spectregnosis/Eternal & Eldritch blog:

https://www.tumblr.com/blog/view/spectregnosis

Nik Strange/Sigils by Strange blog:

https://www.tumblr.com/blog/view/strangesigils/646673261549289472

Children of Artemis:

http://witchcraft.org/secondlife/
https://witchcraft.org/witchcraft-wicca-magazine/

In Second Life:

Ellie Fitzpatrick's House:

http://maps.secondlife.com/secondlife/Hollyleaf/71/188/43

Broom Closet:

http://maps.secondlife.com/secondlife/Livingtree/138/157/25

WitchesFest Mall:

http://maps.secondlife.com/secondlife/witchfest/85/114/701

Westwytch:

http://maps.secondlife.com/secondlife/Satori/45/123/802

Tools:

QR Code Generators:

https://www.visualead.com/qr-code-generator
https://goqr.me
https://www.qrstuff.com

Apps:

There are tons of them out there, these days, some better made than others, and some more useful. Ask your witchy friends what they're using…start some conversations of your own!

These are the apps I've found helpful, personally, and ones I own. Basically, if you hit me up on Tumblr or Reddit and asked me what I use or have, this is what I'd tell you. FYI, none of these people/apps are paying me to say nice things about them, or know that I'm reviewing them in the cases where I'm not so fond of them. Some of them are paid, some are free. Don't take me telling you what I think of these as recommendations, necessarily. Earlier I mentioned I have no fondness for Book of Shadows apps, right? You might have one and love it. I'm just one witch among many, and my opinion and $5.25 will get me a cup of coffee, you know? I'm merely sharing my experience, and if you see something you want to check out, awesome!

One last thought before I dive in: A conversation I'd dearly love to have? A group of witches that includes designers, programmers, coders, and users where we could brainstorm what we want and what is possible in witchy apps.

Labyrinthos

The Labyrinthos app is put out by the Labyrinthos Academy (labyrinthos.co) They offer classes to learn how to read/use Tarot cards, as well as having a store where you can purchase supplies (such as a new tarot deck) if you're looking for one. The app has a game that you can play (or not, if you don't want) that teaches you how to read tarot. It also will do multiple spreads, and read them, and gives you a way to save them, as well as look for trends over time.

Soul Wisdom Oracle Cards by Sundara Fawn

I have them, but truthfully, I don't use them often. Personally, I just didn't bond with them. This deck is made by the same folks that brought you the Goddess Enchantment Oracle Cards. There's a free version, and a paid version. With the paid version, you get to see all 48 cards.

Wisdom of the Oracle Cards, by Collette Baron-Reid:

This is actually one of my favorite decks. I love them! They're beautiful, I love their artwork throughout, and the deck doesn't ignore symbolism. The guidebook is actually helpful. This is one deck that I have as an app where I have the cards in real life, also.

Starseed Oracle by Rebecca Campbell

I like them; Keep in mind, they're a paid deck that has a week trial, and then you buy them, or don't. I liked them enough to purchase them. They have a bunch of spreads, and are really good at telling me what I should be focusing on next.

Runic Formulas

I got it because my sister loves her runes, makes her own sets, and they speak to her. They still don't really speak to me, but when I got them initially, I thought they would be for divination - but it's actually meant for doing magic with runes. The app, which is free, will drive you bonkers with ads. If you're into runes, it'll be worth the $2.99 to not have to deal with them. The how to use section was good, and they're categorized nicely.

Digital Runes

These are the ones I got for divination, that still don't really speak to me. It's pretty, and you face your phone down after you ask your question, and turn it back over when it buzzes at you. The free version of the app allows you to cast one and three rune casts, twice a day. If you want a Celtic Cross spread, you need to subscribe (I didn't.)

Goddess Enchantment Oracle Cards by Carrie Kirkpatrick

Has ancient goddesses, and 1, 3 and 5 card spreads, as well as the opportunity to save your reading. The free version shows you 12 cards. With the paid version, you get to see all 48 cards.

Saged: Sacred Space

I love it. It's a community, and parts of it, like some of their premium classes, are pay as you go; There are plenty of free offerings as well. They have a Premium version as well that includes live ceremonies, live and on-demand classes, and Masterclasses. I love that when you open it, it reminds you that you're entering sacred space. I could see that this could be useful to many, free version or paid. This would be especially true if you're a solitary, or you're in a broom closet.

Sigil Music v 1.0

Another one I got because I was curious. When you open it, there's a spot at the bottom to enter your intent; then you hit the "create sigil" button. It turns your sigil into music, and allows you to change the tempo, and octave. I don't use it much. My husband, a musician, thought it was interesting.

Stone: Crystals, Illuminated

This is awesome for identifying rocks; but what's even more amazing is that you can search by the stone's look…or by a name, or best yet, by what the stone's properties are - like, "joy" or "creativity". It's an awesome little tool.

Moon Phases and Lunar Calendar

I have it, and use it. It gives me more information than I know how to use at this point, truthfully.

Moonly

I have it, but can't say I use it much. There are free parts to it, but to unlock the whole thing, its pay by month, year, or lifetime.

Insight Timer - Meditation App

It's similar to Calm, with guided meditations for beginners, including ones that are meant to go to sleep listening to, and they have courses as well. I have it, and I should use it more.

Chakra Pro

I like it, and don't use it often enough. It allows you to work through your chakras, and set a timer for your meditation, as well as allowing you to balance between the tone for the particular chakra, and nature sounds, such as waves, drums, thunderstorms…

Pentacaster (free)

It basically traces a pentagram, which you're supposed to meditate on or cast a spell to. I own it but removed it from my phone. I was curious, and it was pretty useless.

IOvilus

I questioned whether I should even list having it, lol. I figured it would be good for a laugh, if nothing else. It's a ghost hunting app, and your guess is as good as mine whether it works. Personally, I don't think so, and I believe in spirits and know they hang around. Maybe there are ghost communication apps that work, but I think this one's more a game. It told me a ghost whispered "bacon" at me. Really.

Time Nomad

Actually, I recently downloaded this, and I'm still learning it, but if you're into astrology, this could be helpful. I hope it will be, anyway! I'm taking a class, and I got it because of the class.

Everyday Witch Tarot by Deborah Blake

It's pretty, and it's convenient. I can quickly pull a card of the day, or a single card, or have a 12-card spread, if I want. It has a journal feature so if I want to write about it, I can. I tend to use this one when I'm short on time. I have Robin Wood's tarot deck at home, and the Hanson-Roberts deck at my desk, but if I'm on the go, this one's my go-to.

Ref Guide for Essential Oils:

It's not one of those things I use all the time, but as a reference, it's a good one. It tells you what oils are good for what, and talks about plants the oils are made from. If you want more -like how to take them as supplements, use them as personal care in your own products, and recipes for creating your own blends - then it becomes $9.99 a year.

The Work

Based on the work of Byron Katie, (heard about her in one of Carolyn Elliott's classes) I've found this to be really helpful in dealing with my own attitudes and issues, as it forces you to ask yourself hard questions. After you get your Tarot deck yelling at you...there's this app, to help you sort it all out....

Free Candle

I've had this for ages, and I use it...particularly when I'm in a hotel somewhere. They tend to frown on actual lit candles, being, you know...a fire hazard? Folks have argued that they don't feel "real," and I agree. They don't. If I'm at home in my own spaces, I wouldn't NEED a candle app, anyway. I just light a candle. But if you want at least the symbolism of a candle in a working, and you can't use one for whatever reason, this is better than nothing. After COVID, many folks' living situations have changed, and they may not even have the luxury of an altar to put a candle on, or their host or new roomies might frown on real candles, particularly if they have pets. This can go a long way for being such a little app. Touch the wick to light it, and you blow it out like a real candle (it uses your mic).

Picture This

It's good for identifying local plants you come across. I tried the free trial, but didn't think I'd use it enough to warrant paying $30/year. If I decide I'm going to be foraging plants for my own use, I'll get it, y'know? (I forgot it was still on my phone)

The Divine Feminine Oracle by Meggan Watterson

They're pretty, and I use them to meditate more than anything. I like to pull a card for the soul-voice meditation, and see where it takes me, and what it has to say that I need to hear.

Freedan (virtual altar app)

It's nice; you can put in your own photos, light candles, ring a bell, and you have multiple options of different altars to suit you. My only issue with it is that it's not ***my*** altar. Several of the altars can have photos added to them, which is nice; It makes sense, as the more personal an altar is, the more it will mean to you, and the better it will work for you. I have Freedan, I tried it, and I still prefer a picture of my own altar...but that's my personal preference.

Photo Gallery:

Chapter One: Of Sacred Spaces and Places

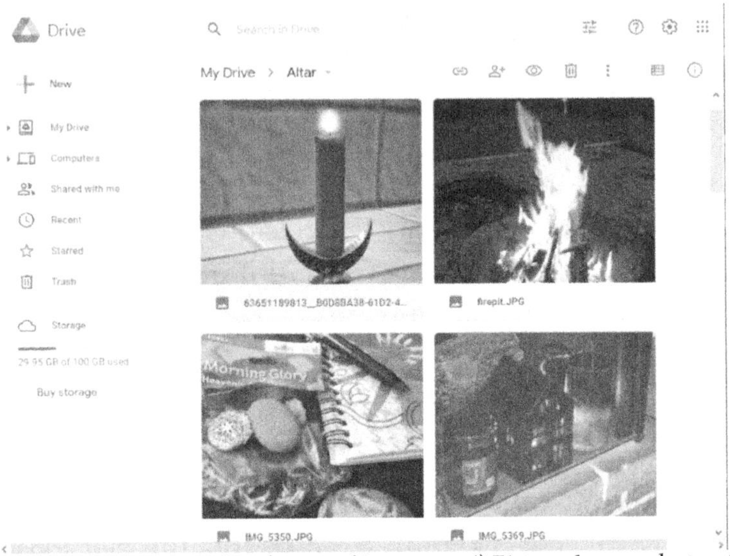

#1: My Google Drive altar (some of it, anyway!) Pictured are a photo of a fire in my fire pit, a spell candle in Second Life, candles burning behind the fireplace screen, and a photo of a spell.

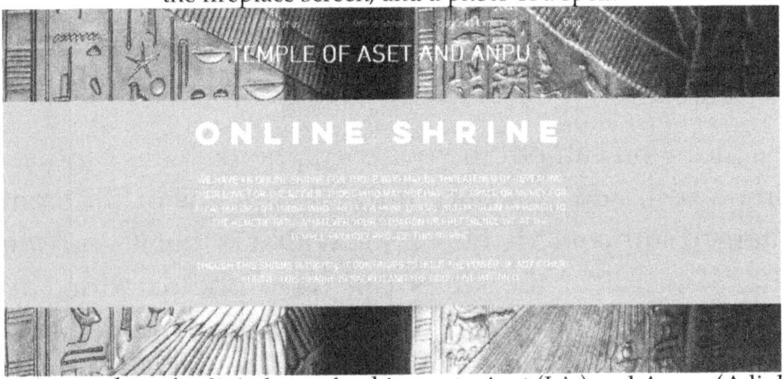

#2: Screenshot of a digital temple, this one to Aset (Isis) and Anpu. (A link to them is available in Appendix 3/Tools, Apps & Helps)

Chapter Two: Altars and Symbolism

#3: My physical altar

#4: My altar in Second Life, with statues and Book of Shadows. The woman in the photo is my mom (deceased).

#5: Closer up picture of my Second Life altar. The woman in the picture is my paternal grandmother, (deceased) also a witch. I use my virtual photoframe to connect to my ancestors.

#6: Louie-Baba-Booey, my altar rock (prefers male pronouns)

Chapter Three: What of Our Tools?

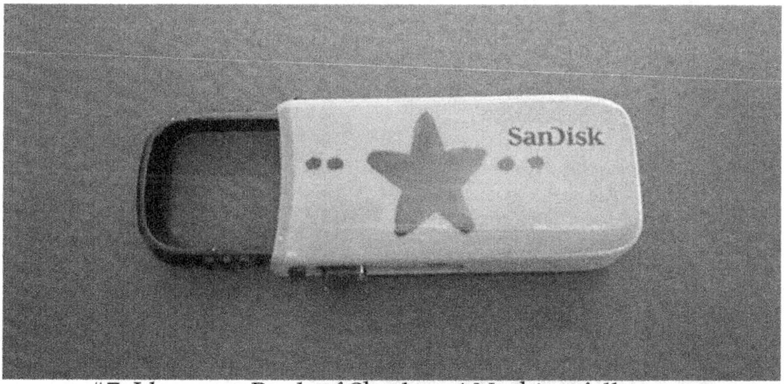

#7: I love my Book of Shadows! Nothing falls out now.

Chapter Four: Virtual Spell Exemplar

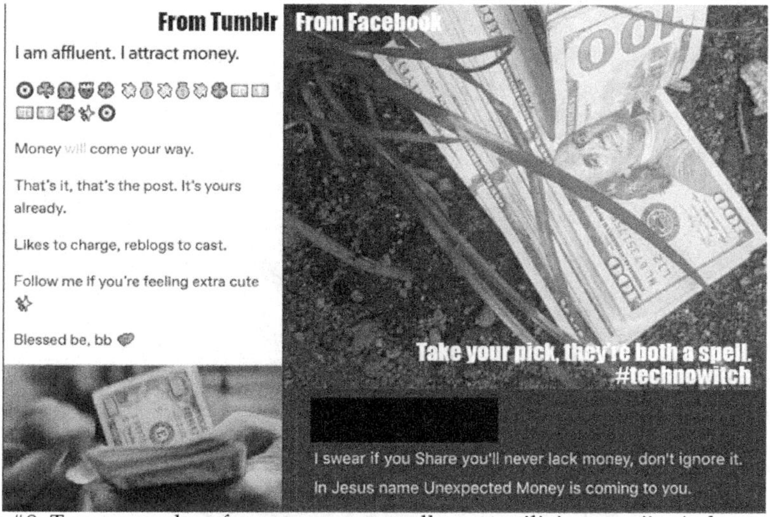

#8: Two examples of money meme spells, one utilizing emojis. At least the emoji spell on the left is obviously a spell…but they both are meant to do the same exact thing.

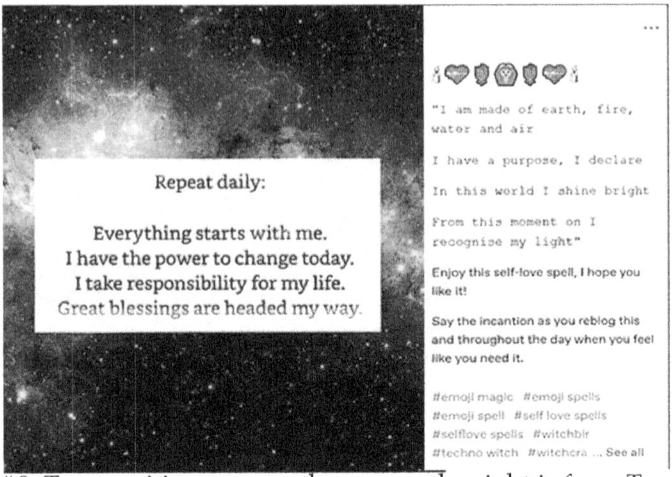

#9: Two positive memes; the one on the right is from Tumblr.
The one on the left, from Facebook.
#10 (below) Share to recast?

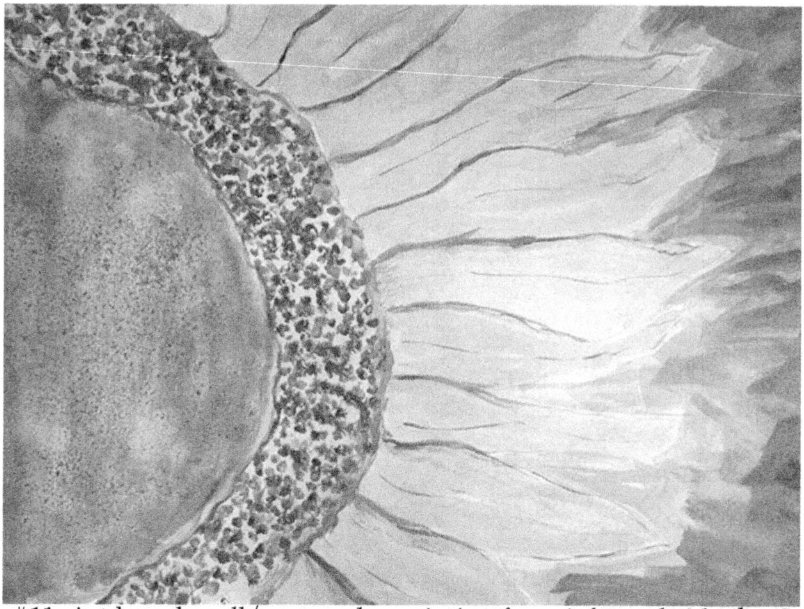

#11: Art-based spell / watercolor painting for grief to subside that I painted. Hangs in my kitchen as a reminder.

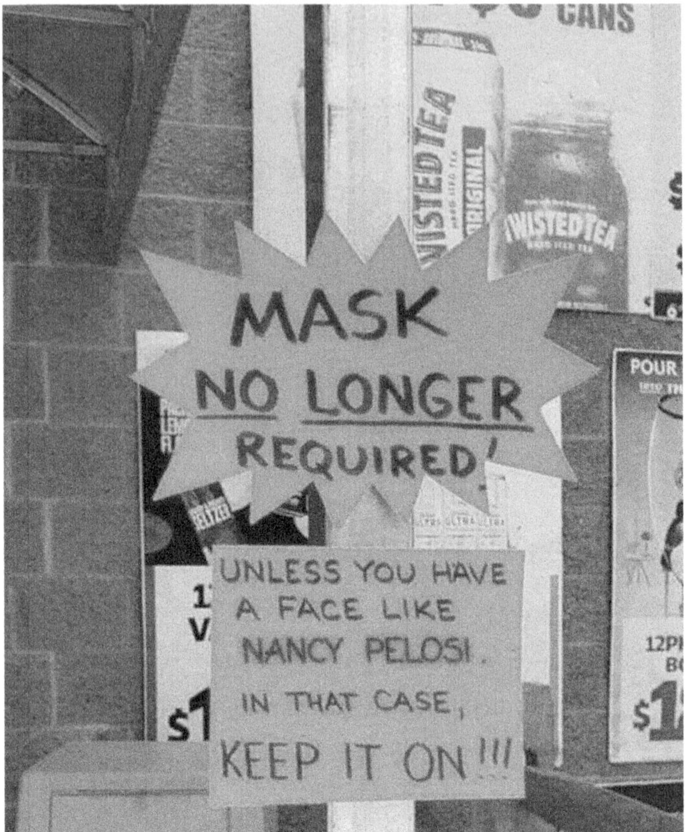

#12: A political meme. On the door, it's an opinion. On Facebook, is it still merely an opinion, a joke that depending on your politics is or isn't funny, or is it magically meant to stir up angst among random people who prefer us at odds for their own reasons? Or…is it some of all of the above?

Chapter Five: Unexpected Magic

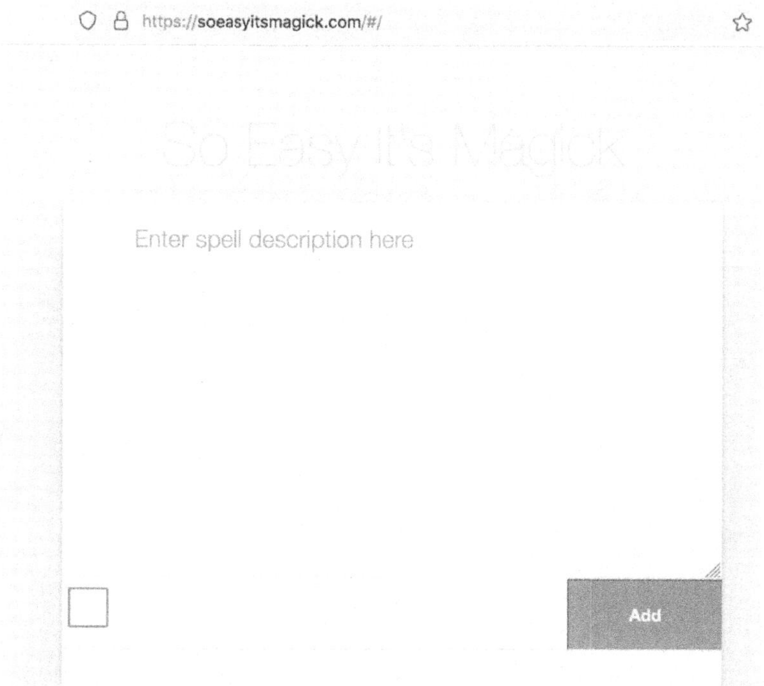

#13: Screenshot of the soeasyitsmagick website.

https://soeasyitsmagick.com/#/

1
10
11
100
101
110
111
I cast a spell with Mercury that people would be o
pen to the materials presented in TechnoWitch, a
nd be aware of the importance of understanding t
hat social media is powered by intent and symboli
sm.
XaTuring, I ask you to reinforce my spell.
111
110
101
100
11
10
1
0

1 item left All Active Completed Clear completed (1)

#14: Screenshot of the spell I put into it. Found out later the next day that XaTuring is the name of the 'God of the Internet' deity some folks are trying to create, and not a programming language. Hopefully, XaTuring and Mercury are on friendly terms. Thanks to AnthropicPrinciple42 for permission to include his program.

SECOND LIFE:

#15 & 16: Ellie at an Egyptian temple. Whatever you're into, you'll find it in Second Life.

#17: (above) Spell candle from my altar. Type the spell in the box, and when you light the candle, spell activates..

#18: (below) Pictures (RL & SL) of Robin "Sojourner" Wood from SL memorial service.

#19: "The Broom Closet" in Livingtree.

#20: Ellie at The Broom Closet.

#21: Artemis Tavern

#22: Sacred Cauldron Library – it's gorgeous!

#23: Main meeting area at WestWytch

#24: One of the areas in Westwytch.

#25 Westwytch.

26: If you try Second Life, visit Helping Haven Visitor Center. There are friendly folks who can help you!

<u>WORLD OF WARCRAFT:</u>

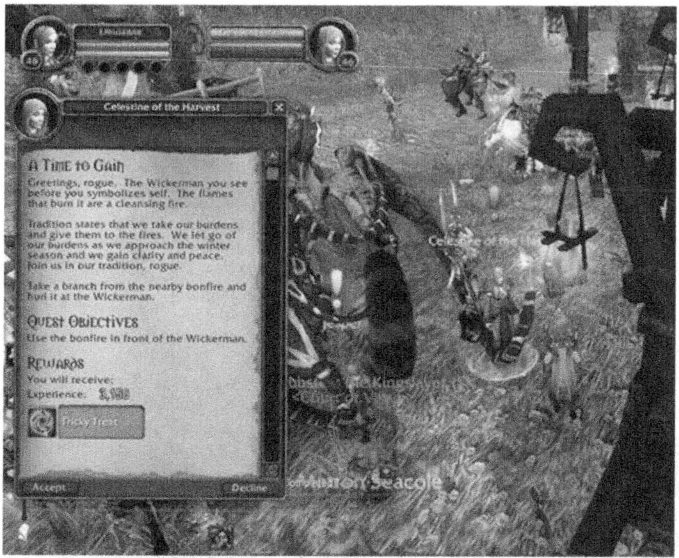

#27: Warcraft Quest from Hallow's End

#28: Throwing my cares onto the Wicker Man effigy as it burns

#29: Moon Festival in Moonglade, honoring ancestors.

#30: At a shrine to the ancestors, at the "secret aerie" in Pandaria.

PHOTOS FROM OTHER GAMES:

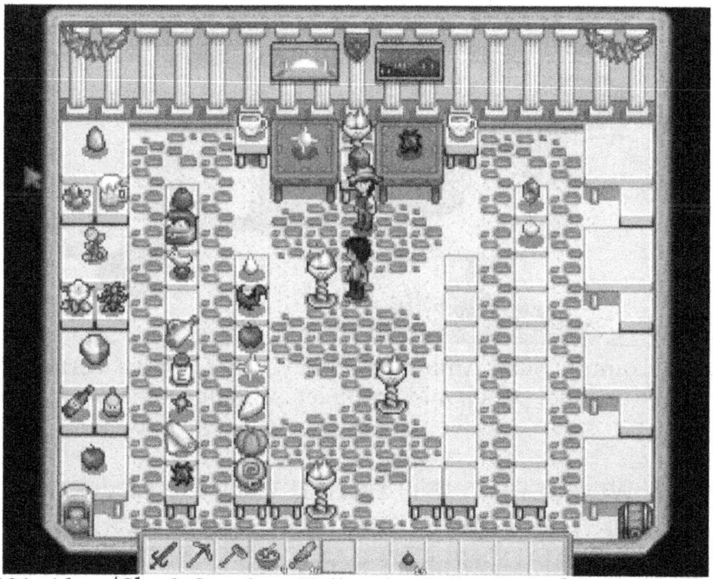

#31: Altar/Shed, Stardew Valley, by Selkie Tidechanger & The Cowboy

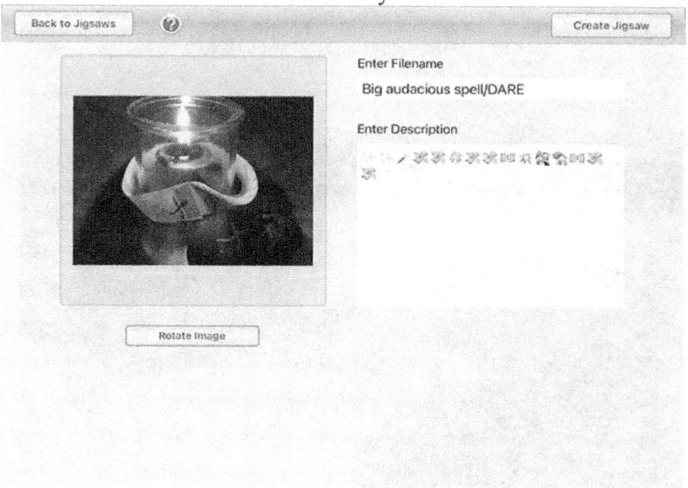

#32: Custom puzzle in Astra Jigsaw. Photo is of a spell for a class; the description is an emoji spell. Separating the puzzle and then putting it together again reinforces the traditional spell.

#33: Example of an Ambiguous Image, By Gustave Verbeek (original panel) and Tverbeek., Public Domain, https://commons.wikimedia.org/w/index.php?curid=57690761

#34: It's all about intent.

Now that you're looking for something else to read, why not check out my other book? It's fictional, called "The Magic Within Us," and it's part of the 'Magic Abounds' Trilogy.

https://www.amazon.com/Magic-Within-Us-Warren-ebook/dp/B0849ZF63K/ref=sr_1_1?dchild=1&keywords=ee+warren&qid=1631849382&sr=8-1

Book Two "The Magic Around Us" will be released in February 2022, and "The Power Within Us" will be released in time for Christmas 2022!

TO THE READER:

Thanks so much for taking the time to read TechnoWitch.

Hopefully, you found it helpful, or something that you read resonated with you, or you found some awesome things you want to try. Or - maybe you realized that something you've been doing for a couple of years now falls into the category of technomagic.

I would sincerely appreciate if you would leave a review for it! It really does make a difference.

ABOUT THE AUTHOR:

This is Ellie Warren's second book. The first one was fictional, called "The Magic Within Us," with the next two books in the trilogy coming out in 2022. She's a kitchen witch, who loves to cook, and paint. When she's not writing, cooking, or painting, she can be found on Blizzard's World of Warcraft, or in Second Life. If you want to know more about Ellie's current projects, be sure to check out her site: http://www.homehearthmagic.com

Made in the USA
Monee, IL
04 November 2026

67379455R00085